Praise for *Untangling Plastics*

"Gedert's book is one of the first to make the critical connection between plastics and their climate impacts, in large part owing to the shared root cause of both crises—fossil fuels. As someone who spent his career in the recycling industry, he also rightly underscores that recycling—be it mechanical or chemical—sounds virtuous but isn't when it comes to plastic. Kudos for plainly saying that by their very nature, fossil fuel plastics 'have no place in the circular economy.' We should all heed the specific actions he lays out, from replacing one's plastic spatula to ending fossil fuel subsidies."

—HOLLY KAUFMAN, director and co-founder,
The Plastics & Climate Project; senior fellow,
World Resources Institute

"In *Untangling Plastics*, Bob Gedert offers a detailed one-stop shop for what we must know and what we must do to turn back the clock on plastic dependence. By exposing the recycling farce, the toxic threat, the LCA blind spot, and the climate connection, Gedert illustrates the threat of plastics' growth trajectory and what it means for the human condition. Importantly, he gets down to brass tacks in laying out the solution. *Untangling Plastics* is an invaluable user guide to saving our plastic-plagued planet."

—JACKIE SAVITZ, former chief policy director,
Oceana; career plastics campaigner; scientist
with training in marine biology and toxicology

"Bob Gedert's *Untangling Plastics* leans on personal experience and science-backed research to thoughtfully unravel the toxic, intertwined issues of fossil fuel extraction, plastic production, plastic pollution, and the climate crisis; Gedert importantly shows us that it is only possible to address all of these aspects of the issue together—plastic pollution cannot be solved in isolation."

—DIANNA COHEN, co-founder and CEO,
Plastic Pollution Coalition

"Bob Gedert's *Untangling Plastics* is a compelling and urgent examination of how plastics contribute significantly to the climate crisis—not just as waste but across their entire lifecycle rooted in fossil fuels. Drawing on his leadership in the zero waste and recycling movement, Gedert exposes the myths of plastic recycling and the deceptive practices of the plastics industry. The book calls for a precautionary approach to safeguard future generations and repositions plastic as a moral issue, not just a policy one.

More than a critique, it's a call to action—for policymakers, advocates, and especially people of faith—to embrace values of justice, humility, and care for creation. With practical steps and a hopeful vision for a Post-Plastic Age, Gedert offers both a prophetic challenge and a practical road map toward a healthier planet and a more just society."

—JAY V BASSETT, retired EPA; former manager
and national advisor, Sustainable Materials
Management Southeast EPA Program;
steering committee member, The Roman
Catholic Archdiocese of Atlanta's Care
for Creation: Laudato Si' Initiative

"*Untangling Plastics* by Bob Gedert, one of the country's most distinguished solid waste and recycling managers, delivers a critically important message to the general public, government and industry officials, and zero waste activists alike. The book lays out the connections between plastic pollution and the direct threats to people and nature. Yet, there can be life on Earth without plastic pollution. Gedert calls for a 'hard stop, now' and then puts forward a clear pathway for collective and individual actions to achieve it. This is a must-read for those working for a fair economy and a sound environment."

—**NEIL SELDMAN, PHD,** co-founder, Zero Waste USA; co-founder and board member, Zero Waste International Alliance

"Whether you're new to the plastic-free movement or one of its most active adherents, Gedert's treatise on the subject of plastic pollution is indispensable."

—**KRISTINE KUBAT,** executive director, Recycle Hawaii; founder, Play Without Plastic

"*Untangling Plastics* is a deeply researched investigation of plastics' impact on the environment and the health effects on humans. Bob notes more than one hundred actions readers can take to eliminate the harms of plastics, including the advocacy of placing extended producer responsibility on the shoulders of the plastic manufacturers. As Bob notes, the precautionary principle of 'do no harm' must be applied to the production of plastics."

—**HEIDI SANBORN,** executive director, NSAC

UNTANGLING PLASTICS

The Missing Link in Mitigating Climate Change

BOB GEDERT

RIVER GROVE
BOOKS

Published by River Grove Books
Austin, TX
www.rivergrovebooks.com

Distributed by River Grove Books

Design and composition by Greenleaf Book Group
Cover design by Greenleaf Book Group
Cover images used under license from ©Shutterstock.com and ©stock.adobe.com

Publisher's Cataloging-in-Publication data is available.

Print ISBN: 978-1-966629-63-4

eBook ISBN: 978-1-966629-64-1

First Edition

*To my wife, Kathy, for her love and patience while
giving me time and encouragement to write this book.*

*To my former college students, who inspired me
to take action toward addressing climate change.*

*To my grandfather, who taught me respect
for our common home, Mother Earth.*

"If it can't be reduced, reused, repaired, rebuilt, refurbished, refinished, resold, recycled or composted, then it should be restricted, redesigned or removed from production."

—BERKELEY ECOLOGY CENTER

CONTENTS

FOREWORD

BOB GEDERT is one of the thought leaders of the U.S. recycling and Zero Waste movements. He is grounded by his practical experience working for the state government in Indiana; local governments in Fresno, California, and Austin, Texas; as executive director of the California Resource Recovery Association; and as a consultant to major cities throughout the country, including New York City and Boston, Massachusetts. He began working on this book while he was president of the National Recycling Coalition (NRC), although this book has not been adopted as the official policy of the NRC.

Bob provides an essential public policy perspective on plastics. He recognizes how challenging plastics have been to recycle throughout his career and that only a few types of the thousands of plastics that exist are currently being recycled. Most importantly, he highlights the impacts of plastics on the environment, climate change, and human health. He notes how recycling plastics is not sustainable for most plastics and that plastics production and consumption need to be reduced to decrease the significant impacts of plastics. He then methodically unlocks keys to phasing out plastics, providing a clear path forward for readers to follow as individuals and as policy and program advocates. He proposes five specific actions regarding plastics in our environment and their impacts on human health:

1. Prevention of harm: Taking preventive action in the face of uncertainty (e.g., avoidance of CO_2 releases into the environment). This prevention is best achieved through the reduction and avoidance of the use of plastics.

2. Burden of proof: Shifting the burden of proof to the advocates of a harmful activity (e.g., production of fossil fuel products). The consumer should not be placed with the burden of proving harm; the burden of proof that a product is safe should be shifted to the producer.

3. Innovation of alternatives: Exploring a wide range of alternatives to harmful actions through innovation, such as creating new replacement products through governmental grants and business innovation centers.

4. Intentionality: Exploring—with a whole intent—to create a new world where humans live in closer harmony with the environment.

5. Inclusion: Involving in decision-making processes all those affected, including the disenfranchised poor and those whose health is affected by the fossil-fuel and plastic-making industries.

Bob asks, "Are we being fed a plastic diet beyond our will?" He suggests, "It is time to act. Individual, collective, governmental, and global action are needed to turn the tide against the plastics invasion."

This book will become an invaluable resource to provide direction for all those currently working on a global plastics treaty and local, state, and federal plastics policies. This authoritative work copiously links to many other resources that advocates can cite to support proposed policies.

—**GARY LISS,** vice president, Zero Waste USA and
NRC Lifetime Honorary Board member

PREFACE

My earliest memories in life taught me patience, perseverance, and appreciation for the support of friends and family. I was born with a "tied tongue," contributing to a heavy stutter. As I attempted to speak, nearly everyone had difficulty understanding me, except my mother, who exhibited great patience and care. To overcome this speech impairment, I attended weekly speech lessons from the time I was six years old through age seventeen. These lessons were life-giving, as I learned to communicate more clearly, suppress the stutter, and gain more command of my vocabulary. I value the time granted by each speech therapist over the years. I may stutter on occasion today. However, it is hardly noticed by those listening to me.

As I struggled to be understood, I often felt isolated from the busy world around me. I found comfort in the local library and became a ferocious reader. I enjoyed the humanities and the sciences and studied the world through books. My best friend, Mark, introduced to me the kindness of friendship. My grandfather also graced my life by introducing me to nature on his family farm and in the backwoods around the property.

The isolation of my childhood played a large role in my view of the world. At first, I viewed my handicap as an adversity to overcome. Through time, I learned that my speech impediment gave me a unique opportunity to view the world in a different manner than my siblings and friends. I was an observer in a world in turmoil, out of sync with its natural order. Through my interest in the natural world, my studies in the library, and my interactions with my grandfather, I became astutely concerned about how humans are directly affecting the balance of nature.

In my turbulent teenage years of rebellion, my grandfather listened to me. As a farmer, he lived and cared for the land he farmed. He played philosopher and counselor to me in his calm and grandfatherly manner. In his well-worn living room chair, smoking his pipe, he listened to my anger at how Earth was mistreated, with pollution seeping into Lake Erie. As we walked the woods behind the farmland, he taught me that an acorn grows into a mighty oak tree when cared for and nurtured. He taught me that nature's growth depends on the previous generation's care for its development and our commitment to do no harm. He also taught me to care for Earth as our collective home.

Grandfather showed me the beauty of fertile soil and the value of crop rotation long before the science books praised "green" farming techniques. Over time, I grew to value Earth and its natural means of recycling and composting "discards." "Waste not, want not" was the frugal lifestyle of past generations, modeled after nature. This respect for nature's renewal and composting cycles grew significantly in meaning to me, physically through nature and as a model for human lifestyles.

As I wondered what my place in this world would be, I felt a growing desire to be a change agent actively working to reverse the human harm placed upon the environment. On Earth Day 1975, I participated in a large tree planting activity, and thus begun my environmental activism.

Fast-forward to my forty-five-year career in the world of recycling. I promoted operational efficiencies in recycling in numerous communities across the United States. My approach involved detailed strategic planning processes, bridging recycling best practices and sustainability with local economic development practices. It has been an enriching career, and I have shared my learned experiences through presentations at hundreds of conferences. I have also proudly served as past president of the National Recycling Coalition and was honored to receive the NRC Lifetime Achievement in Recycling Award in 2019.

In my soft retirement, I am an adjunct professor at Xavier University in Cincinnati, Ohio, teaching classes on climate change. As I teach, my students actively engage me in conversation about the need to take action to address the ever-increasing warming of Earth due to greenhouse gas emissions. While their enthusiasm and urgency are well-placed, I have found that the research was weak on innovative actions that college students can engage in.

My students and I have been energized by Greta Thunberg's and other climate disrupters' leadership. Citizen engagement to challenge the status quo is necessary to address the existential threat we are facing. I teach the profiles of many climate voices from the last seven decades. Sadly, many of those outspoken voices are ignored by decision-makers and those in power.

Yet it cannot be said that they are unaware of the problem. Nearly two hundred nations meet each November to discuss the status of climate issues. The Conference of Parties (COP) meetings offer a platform for global discussions on national contributions and commitments to lower emissions. In addition, the Intergovernmental Panel on Climate Change (IPCC) offers scientific data and advice to the annual COP gatherings.

As I read these IPCC reports and summarize them for my classes, the discussion of this data inevitably leads to exploring how society can eliminate reliance on fossil fuels—the direct, human-made cause of climate change. A 2023 IPCC report notes, "Limiting human-caused global warming requires net zero CO_2 emissions. Cumulative carbon emissions until reaching net-zero CO_2 emissions and greenhouse gas emission reductions this decade largely determine whether warming can be limited to 1.5°C or 2°C."[1]

In translation, this statement—"requires net zero CO_2 emissions"— requires all nations to eliminate the burning of fossil fuels (which emits CO_2 emissions) and take other measures to replace the need for fossil fuels in transportation methods, the generation of electricity, and the production of plastics. Why? To limit Earth's excessive warming and avoid reaching numerous irreversible tipping points in global climate. Choosing a warmer planet is not an option. Continued greenhouse gas emissions will lead to increasing global warming . . . and every increment of global warming will intensify multiple and concurrent hazards.

That leads to why I wrote this book. Many books have been written on the need to transform the transportation system to electric and hydrogen vehicles, convert to renewable sources of electrical energy, and change out our building systems to eliminate reliance on carbon sources. Plastics are also a fossil fuel product that contributes to climate change, yet the connection of plastics to fossil fuels seems untouched. To reach net zero on CO_2 emissions, we need a game plan to migrate away from our reliance on plastics.

This book is written to address that missing connection: plastics' impacts on Planet Earth and how we can better care for our collective home.

ACKNOWLEDGMENTS

My appreciation for the support and assistance from Dianna Cohen and Erica Cirino of the Plastic Pollution Coalition. The Plastic Pollution Coalition is a nonprofit communications and advocacy organization that collaborates with an expansive global alliance of organizations, businesses, and individuals to create a more just, equitable, regenerative world free of plastic pollution and its toxic impacts.

In addition, I appreciate the peer review and support offered by Holly Kaufman, director and co-founder of The Plastics & Climate Project and senior fellow of World Resources Institute. The goal of The Plastics & Climate Project is to help estimate the extent to which plastics and their associated petrochemicals contribute to the global average temperature rise. Holly is a senior fellow at World Resources Institute (WRI), serving in her capacity as thought leader in the nexus between climate change and plastics.

I am grateful to the Greenleaf Book Group editors for their professional editorial support of my manuscript, which has contributed to a collaborative effort in bringing this book to you.

INTRODUCTION

Most people consider plastic consumer items or packaging as beneficial and generally inert or innocuous—not harmful. Yet the plastic products in our household are made from hundreds of chemicals originating from petrochemical factories that also produce oil and gas products. The very companies that are directly impacting the warming of Earth and causing climate change are also producing plastics through the extraction, processing, and production of petrochemicals. What is the relationship between plastics and climate change? How are consumers of plastics complicit in the complicated web of environmental climate impacts? How can individuals change their consumer practices to slow the effects of climate change? These are all questions I hope to address in this book.

All forms of synthetic plastics are made from fossil fuels, and every step of the fossil fuel trail impacts the climate and degrades Earth's environment—from the extraction of minerals and gas; to transport, processing, and refining; to the manufacturing of plastics. Then there are environmental and climate impacts in the packaging and sales of plastic products, consumer usage and disposal habits, and the perpetual impact of plastics on the environment. From the moment of creation through the infinite lifespan of each plastic product, there are continuous environmental impacts that are unmitigated and infinitely ongoing—impacts that ultimately affect the climate and the livability of Planet Earth.

The precautionary principle must be applied here as a framework for public safety and public health. It encourages the pairing of the protection of human health and the environment's safety, where the risks may be difficult to identify or determine with certainty. What we do know is that plastics are extensively harmful to human health and the environment. Applying the Do No Harm precautionary principle would require following the pathway toward eliminating plastics in our lives.

The precautionary principle should not be viewed as an adverse action but rather as a proactive, positive step forward in human relations. It creates a more holistic decision-making process that can create buy-in from citizens who have been harmed.

The plastics industry "has known for more than 30 years that recycling is not an economically or technically feasible plastic waste management solution," as demonstrated in a report from the Center for Climate Integrity (CCI)[1] and numerous other scientific reports. These reports validate what other investigations and news reports have been saying for decades: that the industry has long known that the vast amounts of plastics manufactured cannot be addressed through recycling. It's not a consumer problem; it's a plastics problem. These principles are explored in this book as I do a deep dive into the various harms plastics exhibit.

The starting point for this exploration in the world of plastics begins with the backdrop of our ability to harm—or not harm— Planet Earth, home to eight billion people, two million species of animals, and more than half a million plant species. Through human innovation and intentionality, the decision is to harm or to not harm, but often we don't see the choice.

We then take a journey into the basic definitions and components of plastics and a detailed study of the harms to the climate, the environment, and to humans caused by our collective use of plastics. This is important to understand how we as a society reached the decision point that we collectively stand at today—a decision to continue our current use of plastics and further harm Earth and humans or take a corrective course of action.

We have had voices in the wilderness offering warnings about the warming of the planet and concerns of excessive consumption of plastics. Yet society moved on with its love affair with plastics, until they have become so ubiquitous that you can spot at least a dozen plastic items in any room you walk into. We live on a very fragile planet, where the balance of nature is out of sync, causing ripple effects as we reach certain climate tipping points.

The optimist in me offers the thesis that if this balance was disrupted by humans, it can be reversed and repaired by humans. However, the distress signals from Earth offer a strong sense of urgency. Thus, I offer six societal action steps to address the cause, and multiple actions that individuals can take to assist in the reversal of human-induced climate change.

The following is a quick walk-through of how these themes unfold chapter by chapter.

Chapter 1 explores the concepts of the wasteful practices of our society, the Do No Harm concept with the precautionary principle, actions with intentionality, and the intersection of plastics with climate change. Chapter 2 helps us understand plastics' origins, definitions, and the various "plastic ages" throughout the past century. Chapter 3 dives into the technical definitions of plastic resins and the history of the plastic recycling symbol—the three chasing arrows with a number in the center. A call to action is requested regarding the resin coding system.

Chapter 4 offers a deep dive into plastics' impacts on humans and the environment. Here, we learn about the chemical additives in plastics that affect our health, known as forever chemicals. There is also discussion about plastics in the natural environment and their effects on aquatic life. Finally, microplastics are defined and described in the human and natural environment.

Chapter 5 ties the creation of plastics to climate change, including the extraction, transport, refining, manufacture, and use of plastic products and packaging. It also discusses the early voices of concern that have been dismissed. Chapter 6 discusses the challenges of recycling plastics, including definitions and industry shifts in direction. The critical point is that the recycling industry cannot solve the plastics problem.

Chapter 7 shifts our mindset toward a fragile Earth, an inefficient means to produce plastic products, and a challenge to the legality of harming the population for profit. There is an urgent call for action to protect Planet Earth. Chapter 8 takes us into a paradigm shift: a contrarian viewpoint. The discussion leads toward rejection of compromise and toward patience and innovation. We must challenge conspicuous consumption and rethink our current lifestyle around plastic. I challenge the traditional definition of sustainability as economic interests override care for our common home.

Chapter 9 offers a bridge phase-out strategy: six large system steps necessary toward a full phase-out of plastics and a phase-in of environmentally safe substitutes. Chapter 10 calls for individual, collective, and national actions. All the suggested actions are realistic and feasible.

The Epilogue closes the book with a reminder of the Do No Harm philosophy, the need to restore and regenerate Earth from the harm imposed by plastics, and the need for transformative change through intentionality. Our collective actions will lead to hope for a safer, cleaner, more interconnected future: a Post Plastic Age.

Artwork by Pam Longobardi + Drifters Project

NEWER LAOCOÖN (VOICE OF WARNING), 2015

Ocean plastic from Greece, California, Alaska, Hawaii, and Costa Rica, 96" x 89" x 5". This work updates an ancient Greek sculpture of the complex mythology of Laocoön, a priest of Poseidon who warned of the danger of the Trojan horse and was thus punished by an attack of two sea serpents sent by Poseidon that killed his two sons. Symbolizing a "voice of warning," this work gives voice to the ocean, now sending plastic back to us to warn us of its damaging effect and her state of ill health. Here I envision black and white serpents representing a balance evocative of the yin-yang symbol as a necessary next step in humanity's realignment with nature.

—PAM LONGOBARDI

Chapter 1

DO NO HARM

"It is essential to release humanity from the
false fixations of yesterday which seem now to bind it
to a rationale of action leading only to extinction."

—R. BUCKMINSTER FULLER[1]

I had two childhood heroes during my formative years, both shaping my personality and drive to make a difference. First, my grandfather taught me the farmer's ethic to do no harm to Earth, a primary tenet of the precautionary principle. Second, Neil Armstrong, the astronaut, taught me that if a human can walk on the moon, anything is possible when you set your goals toward achieving your dreams. (Hold in mind that thought about shooting for the moon; we'll revisit it again in Chapter 8.)

When I was a teenager, my grandfather was my hero. He was a family farmer in Northwest Ohio. He cared for the land he farmed through rotation of crops, protecting the soil from depletion. As we worked the land and reaped what we sowed, he taught me that plants will grow when cared for and nurtured. He taught me that nature's growth depends on each generation's care of nature's development and our care for Earth as our collective home.

My dad grew up on the farm, tending to its care. Then he moved to the city to raise his family. I am part of the first generation in our family history to live solely in the city—away from nature except for the squirrels and acorns in the neighborhood. Although my living world has been full of concrete and fences, I visited my grandfather on weekends, seeking a refuge from the city concrete jungle.

Through these visits to the farm, I learned to value Earth and the natural means of recycling and composting its "discards." The motto "Waste Not, Want Not" as the frugal lifestyle of past generations, modeled after nature, came to life. This respect for nature's recycling and composting cycles grew significantly in meaning to me, physically through nature and as a model for human lifestyles.

ZERO WASTE

Tradition tells us that each generation rebels against the previous one, wanting to improve on past mistakes. That is as it should be. Yet we carry forward to future generations the high values and lessons learned from the past. One such lasting value is caring for Earth. The multigenerational stewardship of Earth is humans' study of the natural recycling process, where no "waste" remains—all discards of nature are used to regenerate new plant growth.

I am reminded of a speech I was privileged to hear by R. Buckminster Fuller at a 1976 lecture he gave to business leaders in Chicago. He said: "Pollution and waste are nothing but resources we're not harvesting. We allow them to disperse because we've been ignorant of their value. Any waste as an output of a business is an operational inefficiency."

His point to these executives was that their businesses need to operate more efficiently and do away with the allowance of pollution and wasteful practices. Examples he provided included energy waste, water waste, lack of recycling and waste reduction efforts, and overreliance on long-distance transportation supply networks. He also noted that human "wastes" are due to business reliance on productivity measures rather than worker development and leadership skills. I believe that Buckminster Fuller's speech was the forerunner of the clean (green) manufacturing movement.

Consider the possibility of a waste-free office, school building, or home, where we reuse, recycle, or compost every discard. Zero waste is an approach to our human discard stream, where "waste" (discards) is regenerated—given a second life—through reuse, recycling, and composting. As defined by the Zero Waste International Alliance in 2008:

> Zero Waste is . . . the conservation of all resources by means of responsible production, consumption, reuse, and recovery of products, packaging, and materials without burning and with no discharges to land, water, or air that threaten the environment or human health.[2]

Although we distinguish the human world from nature, we are inseparable from the natural world. Humans depend on Earth and the food supply from natural life cycles in order to live. All of nature involves complex and interwoven processes of life; we are part of those cycles, not distinctly different from them.

At present, our wasteful practices are out of sync with what nature teaches us. Zero waste is a descriptive term that applies to the natural cycles of growth, death, and regeneration into new life. We must learn this lesson of nature and work toward a less wasteful lifestyle, practicing the zero waste principle of nature. A new life begins as the little acorn falls from the mighty oak, not far from its trunk. So can our children also embrace the age-old value of "waste not" and integrate zero waste into their growth and life experiences.

GRANDFATHER'S LESSONS

Along with introducing me to the values and lessons of nature, my grandfather taught me that there are four approaches to a resolution when confronted with a contentious problem too big to resolve with simplistic answers. He'd draw four rectangle blocks: one top, one bottom, one left, and one right, like this:

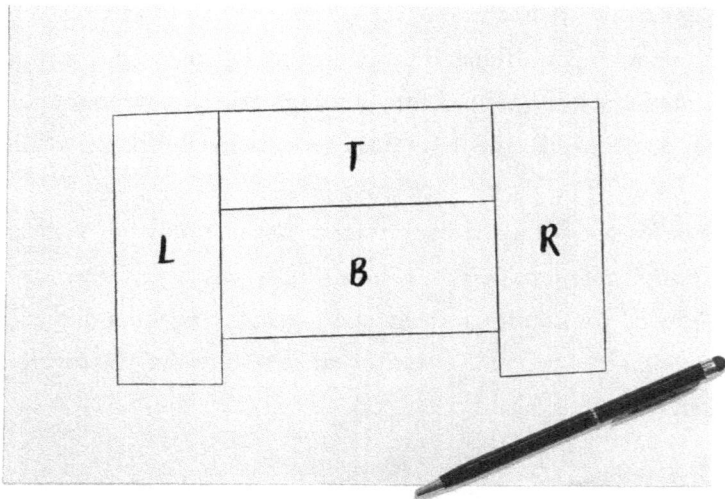

Then he proceeded to explain the meaning of each rectangular cell. The left cell represents my viewpoint, the version of the story where I am right and all others are wrong. Although I may sound righteous and self-assured, that angle of argument will not likely win over many friends and alliances.

The right cell represents the opposite, the opposing viewpoint; I am wrong and the other side is right. Unfortunately, that position will likely be staked out if I take the self-righteous role noted in the left cell, creating the adversarial stalemate such as we find in Congress. Grandfather stated in his slow but confident voice that he had been down that road many times and can show the bruises but no victories.

The bottom cell denotes the compromise position. This fail-safe position is actively selected when all other staked-out positions fail. I give in on some; you give in on some, and we compromise in the middle. Most congressional acts have been passed through this compromise "sausage-making" process. Some praise the art of compromise, but Grandfather said not to stop there, as the easy pathway of compromise often brings dissatisfaction immediately after the agreement is made, with distrust boiling again.

He advised taking the long, rugged path toward the top box: the innovation position. When aiming for innovation, a solution satisfying all parties involved can appear and create mutual trust. However, innovation and trust are slow and require patience—compromise is the fast track most often taken when time is more important than quality.

Some may recognize Grandfather's teaching as being from Stephen Covey's *The 7 Habits of Highly Effective People*,[3] where the innovative cell is described as the Win/Win culture. This is a practical starting point for the discussion of this book. We are seeking a creative, trusting solution through a Win/Win culture and mindset. I ask the reader to leave behind the art of compromise solutions and start up that brilliant mind with the thought processes of innovation—we are on the pathway toward saving Earth through a Win/Win culture, and that journey starts now.

THE INNOVATION POSITION

We live in the age of climate change. This book is intended to address climate change from the perspective of plastics pollution. We have been down the warring paths of "I'm right, you're wrong" and vice versa regarding climate change. Has it provided any progress? Please show me where and how we have made any gains toward the objective of saving our planet from the destructive forces of the climate crisis. The arguments of who is right or wrong have cost us precious time in the war against climate change. In staking out a claim of "I'm right, you're wrong," we effectively display our insecurities about our positions. No ground is gained in this warfare, and meanwhile Earth, its environment, its biosphere, and its people suffer due to our inattention to the needed changes.[4]

Has compromise brought any light toward providing solutions that will slow the warming of Earth's temperature? Compromise is often viewed as a negotiated settlement among warring parties, yet meanwhile climate change continues to fester and grow in its devastating effects. When have you seen public policy from Congress, through the compromise process, bring about the real aggressive change we need to attack the issues head-on at their source? This "temporary balance" between sides has *not* brought about the new and innovative public programs and actions required to address real climate change. What we get through compromise is watered-down actions—yes, an action pathway, but a slow path that does not recognize the short-fused timeline we have to cool this planet from the harm imposed by hundreds of years of CO_2 warming effects.[5]

Climate change is a multi-issue, complicated web of chain reactions resulting in uncontrolled and ever-increasing warming in Earth's atmosphere. Climate Central offers this recent assessment: "The unprecedented rate of global warming observed since the 1950s, due primarily to emissions of heat-trapping gases from burning fossil fuels, is influencing long-term warming trends in each season."[6]

That warming brings increased natural disasters—more intense fires, rising ocean levels, more destructive hurricanes per year, more susceptibility to diseases, changes in the migration of birds and animals, and a faster pace of species extinction.[7] Many more effects of environmental damage on this planet due to accelerated climate change can be listed here. The results are growing exponentially as we waste time through our debates. This book does not debate the reality of climate change—it is real, and it is an ever-increasing threat as Earth's temperature rises. Unfortunately, debates and compromises slow us from finding and implementing the proper solutions we must urgently seek.

We have debated the effects of climate change in the federal government and the U.S. Congress since the 1960s. Naomi Oreskes, a historian of science at Harvard University, offers detailed research that "by the mid-1960s, climate change was already becoming a matter of concern to the federal government. . . . A 1965 report from the National Science Foundation found that the ways humans inadvertently changed the world—through urban development, agriculture, and fossil fuels—were 'becoming of sufficient consequence to affect the weather and climate of large areas and ultimately the entire planet.'"[8]

Carbon dioxide was recognized as a pollutant as early as the 1960s. By 1970, a task force on air pollution commissioned by President Richard Nixon proclaimed in a report that "the greatest consequences of air pollution for man's continued life on earth are its effects on the earth's climate."[9] In 1975, the term *global warming* appeared in print for the first time, in a journal article titled "Climate Change: Are We on the Brink of a Pronounced Global Warming?"[10]

Five decades later, humans are still grappling with climate change's causes and effects, with stalemated discussions in political arenas worldwide.

It's time to explore more innovative platforms and solutions that might help us resolve this stalemate. It's time to move past the decision patterns of the twentieth century and into the twenty-first century. That does not imply higher technology but higher thought processes. It's time to work *cooperatively and collectively* through trust relationships rather than staying forever at odds with one another. It's time to share resources and ideas rather than compete for the top prize. It's time to care for Mother Earth as ordinary people—together, with innovation, through collective modes of innovative thinking.

This book is intended to move in that direction, but it is only one part of the puzzle. There are no silver bullets here—no one has answered the riddle of how to solve the climate crisis. But we can be innovative in working together with many solutions toward the same end goal.

This book explores one of those solutions: the linkage of plastics to climate change and how each of us can take action to make a positive change, thus reducing our environmental impacts. Innovation will take us down that pathway.

INTENTIONALITY

Any list of the biggest, most memorable stories in TV media history is likely to include mankind's first moon landing on July 20, 1969. The unforgettable images of Neil Armstrong and Buzz Aldrin walking on the moon while Michael Collins circled in the command module are iconic. They've become synonymous with American innovation, human achievement, and the impossible made possible. So what happened a half-century ago that sprung this far-reaching scientific advancement? The answer is intentionality.

President John F. Kennedy stated his intent to reach the moon in a visionary speech in 1961: "First, I believe that this nation should commit itself to achieve the goal, before this decade is out, of landing a man on the Moon and returning him safely to Earth."[11]

This commitment turned America toward a then unimaginable goal, with full *intentions* of reaching that goal by the decade's end. Declaring that vision led to the landing on the moon in 1969. In addition, the space program is credited with thousands of scientific advancements in more than five hundred consumer products, including Velcro®,[12] microwavable meals, and the joystick controls we utilize to operate our games.

This American story is a classic example of intentionality—through the actions of scientists who believed in the mission and the visionary goal. It was the same act of intentionality that brought us many modern inventions in medicines, such as penicillin (Sir Alexander Fleming, 1928), the polio vaccine (Dr. Jonas Salk, 1955), and the smallpox vaccine (Edward Jenner, 1796). It was an act of intentionality to ban DDT from production and use (1972) after the publication of *Silent Spring* by Rachel Carson brought to light the deadly effects of the chemical on humans and the environment. Intentionality is at the heart of human progress. Whether we judge that progress to be supportive or harmful is an opinioned and debatable manner in each situation.[13]

We urgently need human imagination, intentionality, and innovation to protect our people, environment, and planet. As discussed in Chapters 8 and 11, it will take visionary goals, new missions, innovations, strong intentions, and a sense of urgency to change the course of the current climate crisis threatening Planet Earth.

DO NO HARM

I began this chapter with a quote from R. Buckminster Fuller regarding past decisions that have led us toward extinction. This fate he warns of becomes a reality only if we stay the course of our past choices and lifestyles. Another critical concept is Do No Harm, embodied in the precautionary principle.

As embraced by the United Nations in 1992, the precautionary principle declares that "where there are threats of serious or irreversible damage, lack of full scientific certainty shall not be used as a reason for postponing cost-effective measures to prevent environmental degradation."[14] The UN statement also declares that "in order to protect the environment, the precautionary approach shall be widely applied by States [Nations] according to their capabilities."[15]

The concept of the precautionary principle dates back centuries. It is expressed well in the Hippocratic Oath—from the classical Greek to our modern American Medical Association's Code of Medical Ethics that doctors and nurses take as they enter their careers. The physician must first "do no harm" in treating a patient when the risks of treatment may be uncertain. Likewise, the precautionary principle applies to climate change, providing the framework for public safety and public health.

I propose five specific action values in the application of the precautionary principle: prevention of harm, burden of proof, innovation of alternatives, intentionality, and inclusion. (I offer more detail for applying these action values to plastics in the Epilogue.) By applying the precautionary principle within these five proactive values, we can advance the protection of humans, the environment, and the climate. As we proceed through this discussion in later chapters, there will be more on the precautionary principle.

PLASTICS, FOSSIL FUELS, AND CLIMATE CHANGE

The climate is impacted directly through the extraction of gas and oil to create plastics through releasing greenhouse gas emissions, including carbon dioxide (CO_2) and methane (CH_4). Throughout the creation and manufacturing of plastics, there are additional releases of CO_2 and CH_4. As plastic products and packaging are transported, marketed, consumed, and disposed of, there are additional releases of CO_2 and CH_4. These greenhouse gas releases impact climate change, human health, and the local environment.

Let us now move forward into Chapter 2 to discuss the progression of plastics into our consumer lives. In Chapter 3, we will explore the interaction of plastics within the environment, and Chapter 4 will introduce the connection of plastics to climate change.

Artwork by Pam Longobardi + Drifters Project

ANTHROPOCENE TIME CAPSULE, 2018

Ocean plastic, including thousands of cigarette butts, plastic beach toys, Mardi Gras beads, flip flops, collected by Fight Dirty Tybee and Tybee Clean Beach volunteers; vitrine; and wood pedestal. Collection of the Telfair Museum Jepson Center and donated to Tybee Island Marine Science Center.

—PAM LONGOBARDI

Chapter 2

PLASTICS: A DEFINITION AND HISTORY

"Perhaps the simplest example is a synthetic plastic,
which unlike natural materials, is not degraded by
biological decay. It therefore persists as rubbish
or is burned—in both cases causing pollution."

—BARRY COMMONER[1]

In nearly all books I have read on plastics, there is a standard reference to the iconic scene in the 1967 movie *The Graduate* where the lead character, a college graduate played by Dustin Hoffman, is given a bit of unsolicited career advice by a businessman: "I just want to say one word to you. Plastics! . . . There's a great future in plastics."[2] The implication was that a career in the world of plastics would be lucrative.

Consider the recent earnings of Dow Chemical, a plastics manufacturer, as one example of wealth in the plastics marketplace. The 2020 third quarter financial report displays an overall assets level of $60 billion (liabilities are charged to those assets). The nine-month net sales, as of September 30, 2020, was $13,175,000 for the "packaging and specialty plastics" sector, representing nearly 50 percent of total company net sales for those three quarters.[3]

If Hoffman's character did invest one hundred dollars in plastics stock in 1967 and allowed the annual dividends to reinvest, using Dow Chemical's numbers as a base reference, the current value of that stock would no doubt be worth hundreds of thousands of dollars. Calculating the exact value is quite difficult, as the stock has "split" five times between 1967 and today, and the stock valuation has seen some roller-coaster years. The point is that the plastics industry has been quite a rich investment over the years, but there is another side to the story—the cost to humans and the natural environment. We will get to that story soon enough, but first, what are plastics?

WHAT ARE PLASTICS?

The first written use of the term *plastic* seems to have been in the 1630s. According to the *Online Etymology Dictionary*, *plastic* is an adjective meaning "capable of shaping or molding a mass of matter," from Latin *plasticus*, from Greek *plastikos,* "fit for molding, capable of being molded into various forms; pertaining to molding."[4]

By current scientific definition, *plastics* are a group of synthetic or natural materials capable of being formed or shaped. Most plastics are polymers made of many repeating units at the molecule level.[5] The feature that makes plastics so interesting for scientists to explore is their ability to be pliable and easy to shape when the material is warmed to a peaceful state and then hardened to a shape that it retains (thermoplastics). Some plastics may even maintain flexibility at room temperature (thermosets).

Today, 99 percent of plastics are synthetic (artificial) and not found in nature. Yet the earlier versions of plastics that were utilized were naturally occurring plastics. This included tar, shellac, tortoiseshell, animal horn, cellulose, amber, and latex from tree sap.[6] Synthetic polymers are made from hydrocarbons and are not naturally occurring but are human-made and *fabricated from fossil fuels*. Simple synthetic hydrocarbons can contain only carbon and hydrogen. However, more complex hydrocarbon polymers may contain carbon, hydrogen, oxygen, fluorine, nitrogen, and chlorine.[7]

That may be the scientific definition of plastics, but you and I, as consumers, see plastics from a different perspective. We see plastics replacing everyday household items, sometimes due to cost savings and sometimes due to the ease of manufacturing the parts from plastic resins instead of the original replaced materials.

The business world sees plastics as having primary attributes: cost-effective manufacturing methods, durability for their weight class, electronic and chemical insulation, and resistance to chemicals and corrosion. Many would claim plastics are water resistant, though we have recently learned of their fragility in ocean water. (More on that later.)

PLASTICS AS SUBSTITUTES FOR OTHER MATERIALS

Plastic has replaced other materials in consumer products and packaging. Plastic parts have replaced metal parts in standard household water plumbing fixtures. Due to the physical softness of plastic valves, these fixtures don't last as long and leak sooner, yet they may come at a lower retail price. Acrylic plastic often replaces everyday household glassware, as it is lighter and more resistant to breakage; however, it scratches easily. Rubber products have been replaced with thermoplastics, adding abrasion and chemical resistance, but they have shorter lifespans.

You will find plastic resin replacements in carpets, microwavable containers, vehicle bumpers, wallpapers, credit cards, and vehicle instrument panels. Most of these items would have originally been constructed of linens, paper, glass, and metals readily found in local manufacturing markets.

Plastic flexible packaging has replaced paper, wax, and foil wraps used in the past. Ceramics and glassware have been replaced with plastic cups, many in disposable single-use format. Silverware is commonly replaced with plasticware. I typically find dessert displays at restaurants replaced with displays of plastic food that look tasty and last indefinitely. For those who like fishing, plastic fishing baits that look real to the fish have replaced fresh bait, and polyester filament has replaced the fishing strings of the past.

I once carried my books to school in the 1960s with paper book covers and corded book straps; today, it is the norm to carry books in backpacks made from high-density polypropene, nylon, polyester, or polyvinyl chloride—all different forms of plastic resins. My Boy Scout backpack was made of cotton canvas and lasted twenty-plus years, yet today's hiking backpacks are made of thermoplastic polyurethane and nylon. Likewise, our warm winter jackets that were once made of cotton, wool, and down are now replaced with a thermoplastic or nylon shell and polyester fleece lining. Winter boots once made of leather and rubber are now made of polyurethane, thermoplastic, polyester, and sometimes faux fur, which is made of polyester fiber. These many and varied substitutions illustrate two points: the replacement of ordinary natural materials like cotton with plastic, and the many plastic resins replacing commonly available materials today.

I could go on for many pages, but you get the idea. Just look around you and discover the plethora of plastic riddled throughout our modern world. Where there is plastic, there has been a substitution for a previously used item. Ask yourself, "What was used before this item was made of plastic? Is plastic an improvement or not?" The answer depends on the situation and is not universally in favor or disfavor of plastics. However, there is one universal truth: *The production of synthetic plastics requires extraction, refinement, production, and transportation of fossil fuels (coal, oil, or gas) to the production facility where plastics are manufactured.*

Think about that. *Plastics are fossil fuel products.* Plastic production requires fossil fuel production. The two are intrinsically linked. This is an essential point to focus on. We will revisit this topic in the following few chapters, but for the moment, the question is this: When did we get hooked on plastics?

A HISTORY OF PLASTICS

Naturally occurring plastics such as tar, shellac, tortoiseshell, animal horn, cellulose, amber, and latex from tree sap have been used for centuries and recorded in written documents dating as far back as the thirteenth century. However, the word *plastic* was not consistently applied.

The first recorded human-made plastic originated from the story of John Wesley Hyatt, who responded to an 1863 newspaper ad offering "ten thousand dollars in gold to anyone who could come up with a suitable alternative for ivory."[8] At that time, ivory came from African elephants to make billiard balls, and the sport's popularity was causing a shortage of ivory (not to mention the extinction of the elephants). A substitute material with the same hardness qualities as ivory was needed so as not to affect the billiards game's quality.

Hyatt invented celluloid, a fabricated form of plastic made from natural elements, including cellulose, nitic acid, and cotton.[9] The new plastic worked well for making billiard balls. With the success of the billiard ball, celluloid became a standard household plastic used to manufacture other items such as combs, brushes, toys, ornaments, table tennis balls, and guitar picks. Few people nowadays remember the time before plastic combs. Celluloid combs were first produced in the late 1800s and became extremely popular due to their low price. Before that, manufactured combs sold in stores were reserved only for the upper class. Plastic combs brought it down to the common social status.

At one point, the movie industry replaced camera film with celluloid. The material lost favor over time due to its flammability and was then replaced with acetate, another type of plastic film. In the 1960s, celluloid gave way to molded fossil fuel plastics, completing the transition from natural materials to synthetic material-based plastics.

Synthetic plastics, made from fossil fuels, were first displayed to the public in 1907 through a product called Bakelite.[10] Invented by Leo Baekeland in a Dupont lab, this substance was a combination of formaldehyde and phenol (a derivative of coal) that was heated and pressurized. Bakelite was the first modern-day type of plastic that could be shaped and molded to industrial standard precision. It could be injection-molded to make industrial, commercial, and consumer products, all from the same feedstock material, which was made primarily from a fossil fuel source: coal. The Bakelite Corporation manufactured a range of products, including Bakelite radios. The possibilities were so exciting at the time that *Time* magazine's September 1924 issue included a photo of Leo Baekeland on the cover and an article describing the many plastic products of the Bakelite Corporation.[11]

The next leap forward in synthetic plastic research also happened in the Dupont labs, where, in 1927, they announced a thermoplastic polymer called nylon. A 1938 press release from Dupont notes the "development of a wholly new synthetic material of hundreds of potential uses, one of which will be of revolutionary importance in fine hosiery."[12] "Nylons" were first presented as a form of women's stockings at the 1939 New York World's Fair. Dupont sold $34 million in nylons in the first two years of commercial sales.[13]

In the 1930s, Dow began producing plastic resins, which would grow to become one of the corporation's major businesses. Its first plastic products were ethyl-cellulose, made in 1935, and polystyrene, made in 1937.[14]

THE WARTIME PLASTICS AGE

Through the 1920s and 1930s, plastics were mainly in a discovery stage of development in scientific labs; most had not yet made it into the stage of functional usage. As necessity is the mother of invention, WWII heralded what many journalists have called the Plastics Age but might be more appropriately labeled the Wartime Plastics Age.

During WWII, nearly all machine shops, auto plants, and extensive production facilities turned toward producing goods for the war effort. Raw material feedstock was needed to feed this production. As essential metals, wood, and rubber became in short supply, salvage drives were established to recover material and reuse and recycle whatever could be found for the war effort. Yet, the war consumed vast amounts of material. The nationwide salvage drives successfully collected large volumes of recycled material but could not supply enough consistent, clean material for the wartime production facilities.

The Allied forces, including the United States, turned to the high production of plastics. Producing synthetic plastics became a wartime priority, as natural materials were in short supply or not obtainable due to shipment disruptions. During WWII, nylon thread plastics were used for parachutes, ropes, body armor, and helmet liners. Plexiglas and acrylics were introduced as an alternative to glass for aircraft windows. Wartime plastic production in the United States increased by 300 percent.[15] A *Time* magazine article noted that because of the war, "plastics have been turned to new uses and the adaptability of plastics demonstrated all over again."[16]

The Manhattan Project turned to Teflon (the brand name for

polytetrafluoroethylene made from carbon and fluorine) for its resistance to gas corrosion. Synthetic rubber was needed for tires, and vinyl polymers (polyvinyl chloride) replaced natural rubber in machine parts. New forms of plastic resins were being explored to substitute for natural material shortages. Production of plastics almost quadrupled during the war—from 213 million pounds in 1939 to 818 million pounds in 1945.[17]

THE JETSONS PLASTICS AGE

After WWII, the second plastics age began to take shape. I label this era the Jetsons Plastics Age because the focus was on the consumer, with a Jetsons cartoon-like appearance and a desire for modern kitchen appliances and furniture. The new consumer products were made of hard, rigid plastics, inexpensive, pigmented in bright colors, and offered new shapes and molded forms. Susan Freinkel, in her classic book *Plastics: A Toxic Love Story*, notes, "In product after product, market after market, plastics challenged traditional materials and won, taking the place of steel in cars, paper and glass in packaging, and wood in furniture."[18] The fascination and new design possibilities were so exciting that various art museums displayed plastic furniture pieces as modern art!

The auto industry had long been interested also in durable, lightweight material substitutes for its heavy steel frames and sheet metal used throughout its history. In 1983, the Pontiac Fiero introduced the first all-plastic chassis in the United States.[19] This chassis was made from reaction injection molding, where reactants are mixed and then injected into a mold. The resulting plastic resin has reduced thermal expansion and more muscular stiffness in its shape. This new plastic molding innovation has evolved over the years, and soon nearly every vehicle on the road had significant quantities of plastic components.

CONSUMER SINGLE-USE PLASTICS

No history of plastics would be complete without some mention of the ubiquitous single-use plastic bottle. The introduction of plastic bottles was slow at first and began in the post-WWII period of plastic experimentation. Food-grade plastic bottles started showing up on grocery shelves in the 1970s, primarily as a substitute for glass to reduce the amount of breakage in shipments and the heavy weight associated with glass bottles and jars. A wide variety of plastic resins were utilized in the production of these plastic containers, including polyvinyl chloride (PVC), vinyl (V), poly-vinylidene chloride (SaranTM), Teflon®, polyethylene terephthalate (PET), high-density polyethylene (HDPE), low-density polyethylene (LDPE), polypropylene (PP), and polystyrene (PS).[20]

The classic PET soda bottle began its birth in 1973, utilizing a black stabilizer cup glued on the bottom of a 2-liter plastic injection molded polyethylene terephthalate (PET) bottle. Although the bottling industry claimed the bottle was "recyclable," recycling it required removing the stabilizer cup from the main body—a laborious task that neither the consumer nor the recycling center could perform. As the general soda industry was resistant to dealing with the issue, the National Recycling Coalition applied for a research grant from the U.S. Environmental Protection Agency (USEPA) Region 5 and contracted with a Midwestern university research lab to redesign the 2-liter bottle. The project's goals were fourfold: a design for full recyclability; a plan to reduce any impacts on manufacturing, soda fill operations, transportation, and sales; a design to reduce any adverse effects on the consumer; and the completion of these tasks within the grant-awarded value of fifty thousand dollars. All four goals were met within ninety days of the start of the project, and the final design was a unibody blow-molded PET plastic resin bottle that met the specs of the bottling and soda industries. The design was open-sourced and made available to all production companies. This final bottle version was recyclable worldwide and began the march toward plastic bottle recycling.

In 1977, the new slim unibody PET single-use bottles hit the market, with Perrier launching their advertising campaign for sparkling carbonated distilled water in the U.S. markets. Other competitors followed. Today, bottled water in single-serving throwaway bottles is the second most popular domestic beverage in the United States.[21] Although these bottles are recyclable, they were initially marketed as disposable for the consumer's convenience.

Euromonitor International's global packaging trends report estimates that more than 583.3 billion plastic bottles were produced worldwide in 2021.[22] The Container Recycling Institute estimates that only 30 percent of plastic bottles get recycled in the United States,[23] and some experts estimate that only 8 percent are recycled worldwide. Note that recycling statistics are generally reported in the United States and Europe, where recycling is more advanced, but it is far more difficult to track the recycling of plastic bottles worldwide. The remainder of these bottles that are not recycled are landfilled (at best), incinerated, littered, or disposed of in the waterways that lead to Earth's lakes and oceans.

A close cousin of the disposable plastic bottle is the disposable plastic straw. Straws became popular at ice-cream parlors for drinking the classic milkshake. The first straw was made of paper and invented by Marvin Stone in 1888.[24] Sometime in the 1960s, the paper straw was replaced with the plastic straw. Today, Americans sip from approximately 500 million plastic straws daily![25] That calculates out to 182.5 billion straws consumed and discarded by Americans annually. As of October 2022, eight U.S. states and several local communities have passed legislation to limit the use of plastic straws in their jurisdictions.[26] In England, where 4.7 billion plastic straws are consumed yearly, a country-wide ban on plastic straws and plastic stirrers took effect in October 2020.[27]

If not banned from production and distribution, where do these straws go after their single use by the consumer? Landfills, incinerators, litter, rivers, streams, lakes, and oceans. It is estimated that 8.3 billion plastic straws litter and damage the world's beaches yearly.[28] Alternative material types for straws being explored are aluminum, stainless steel, bamboo, glass tubes, silicone, paper, and pasta. These alternatives are being tested and marketed to consumers, and I encourage you to try them out for yourself, or go straw-less for waste reduction.

The single-use plastic bag was invented in 1965 by a Swedish company employee, Gustaf Thulin Sten, who designed the manufacturing process of the "T-shirt" plastic bag, named after its shape.[29] Single-use plastic bags were introduced in the United States in 1979. (Yes, there was life before single-use plastic bags!) The new plastic bags gained a significant foothold in 1982, when Safeway and Kroger, two large grocery chains, began offering single-use plastic bags to their customers for free, replacing the traditional practice of the customer bringing in a reusable bag. (The grocers built their bag purchase expenses into their overall food price structure.) If a customer forgot their reusable bag, the prior practice was to use paper bags as a backup. The new approach was to promote plastic bags over both reusable and paper bags. This is an example of a plastic product replacing reusable and recyclable products, thereby changing consumer practices through the guise of convenience.

The plastic bag is made of low-density polyethylene and is primarily utilized in grocery stores, food markets, convenience marts, and retail stores. A single-use plastic bag's estimated "working life" is ten to fifteen minutes, enough time to bring your groceries or purchased products to their destination and unpack. Approximately 500 billion plastic bags are used worldwide each year. That equates to one million bags used every minute.[30] Plastic bags start as fossil fuels and become deadly waste in landfills and the ocean. Birds often mistake shredded plastic bags for food, filling their stomachs with toxic debris. For hungry sea turtles, it's nearly impossible to distinguish between jellyfish and floating plastic shopping bags. Fish eat thousands of tons of plastic a year, transferring it up the food chain to bigger fish and marine mammals.[31]

According to the U.S. International Trade Commission, the 100 billion single-use plastic bags used annually in the United States are made from the estimated equivalent of 439 million gallons of fossil fuels, costing retailers an estimated $4 billion.[32] Rarely do consumers consider the added cost to their groceries as the management recovers the cost of these "free" bags given to customers at checkout at no visible charge.

The end-of-life of these single-use bags has created much debate. Many who argue against the recent municipal bans state they "reuse" the bags as trash can liners or doggy cleanup bags. Yet, that is only a temporary life extension before it ends up in the landfill and is not truly reused. A reusable bag should gain ten to a hundred reuses if designed well and cared for properly.

Customers are encouraged to bring their used single-use plastic bags to recycling barrels at entrances at retail store distributors. Finding reliable sources to track the actual recycling of plastic bags through store collection barrels is difficult, however. I have seen studies stating a range from 1 percent to 10 percent of the bags produced being recycled, without much validation behind the statistics. Curbside recycling programs cannot accept plastic film bags because they can become entangled in the sorting equipment at the material recovery facility (MRF), causing significant downtime to cut out the plastics from the equipment. The remainder of these bags that are not recycled are landfilled (at best) or left to endanger Earth's environment.

Single-use plastic bag bans have recently gained popularity to address the bags' environmental impacts. The bans "stop the bleeding" by prohibiting the distribution of single-use bags or by taxing the bag so that consumers might choose reusable bags instead. A large mixture of bag ban language and policy has come into play, depending on the community debates and the will of the local politicians.

In a 2018 report, the United Nations Environment Programme, developed in partnership with the World Resources Institute, researched plastic bag regulations worldwide and found that at least 127 out of 192 countries reviewed have adopted effective forms of plastic bag regulations. Of those 127 countries, 55 restrict the retail distribution of plastic bags and place restrictions on manufacturing, production, and imports.[33]

Eight states have enacted bag bans within the United States, while one additional state plus the District of Columbia has imposed fees on plastic bags.[34] Four hundred seventy-one local plastic bag ordinances have been adopted across twenty-eight states.[35] Some states have seen negative side effects with banning single-use bags, where retailers began using thicker plastic bags and claiming them as "reusable." Those states are revising their legislation to outlaw thicker plastic bags.[36]

On the other hand, seventeen states have enacted "preemption laws" that prohibit local communities from passing bag bans against single-use plastic bags.[37] (I happen to live in one of these states.) The state-wide preemption laws have received lobbying support from the American Chemistry Council and the American Progressive Bag Alliance, arms of the fossil fuel industry.

THE UNIVERSAL PLASTICS AGE

The explosive use of plastics is not restricted to single-use plastic bags, water and soft drink bottles, nor is the solution toward recycling plastics limited to this small segment of consumer sales. The post-war era of plastic consumer products exploded onto the scene in a display of color and flare, with a shade of newness and appeal for something different. However, as plastic injection molding was a low-cost entry point for new entrepreneurs, the 1970s and onward brought new markets for the plastics industry as they saw a new consumer interest in low-cost consumer products. This third wave of plastics is what I call the Universal Plastics Age simply because plastics now invade every area and aspect of our lives.

Search your home and you will find plastics in every room. In the living room, note the seating: polyester fabric on the sofa and chairs; vinyl plastic covers over seat cushions made of polyurethane foam; polyvinyl chloride or high-density polyethylene leg cups to prevent dents and scratches on the floor. The polyester weave carpet or rug. The high-impact polypropylene lamp base with the "faux silk" nylon lamp shade. The electrical cords are wired in thermoplastic polyester plastic resins. The wall frames may be wood or metal, but chances are at least some of them are polypropylene or low-density polyethylene.

If not granite or stone, the kitchen counters might have a laminate surface made from melamine resin, a form of thermosetting polymer, with the most famous trademark name being Formica. The measuring cups and teaspoons are likely made from low-density polyethylene or high-density polyethylene plastic resin, as are the bowls, cutting boards, and even some dishware. Food storage containers such as Tupperware are made of low-density polyethylene or polypropylene. Some standard kitchen appliances now have plastic components or are entirely made of plastic, such as your plastic coffee maker or the plastic body of your microwave. My new high-quality kitchen cabinets have plastic (not metal) hinges with soft-close features. Common kitchen flooring comprises polyvinyl chloride, vinyl, or rigid polypropylene resin plastic. If made of plastic, the trash and recycling containers are made of polycarbonate plastic.

The bathroom walls may be painted with water-resistant enamel-coated-with-alkyd resins, which are thermoplastic polyester plastics.

The shower curtain may be made of flexible, washable vinyl resin plastic, while the toilet seat may be vinyl resin. Toothpaste tubes, once made of flexible aluminum, are now rolls of thin plastic laminate sheets—usually a combination of different plastics. Shampoo bottles are generally high-density polyethylene plastic, while prescription medicine bottles are polypropylene. The medicine cabinet shelf might be made of hard rigid plastic such as vinyl or polypropylene resin. If a modern prefabricated shower enclosure exists, the walls will be acrylic thermoplastic homopolymer and fiberglass (glass fiber-reinforced thermoplastic). The bathroom flooring may be polyvinyl chloride, vinyl, or rigid polypropylene resin plastic if not stone or tile.

That is a survey of only three rooms in your home. Moving on, consider the sporting goods arena. Nearly every sports ball is replaced with foam plastic to rigid plastic. National Football League rules still require natural leather, but non-NFL-sanctioned footballs are now made of different plastics—sometimes sandwiched layers of synthetic rubber and plastics. However, times are changing; the World Cup of Soccer (the other form of football), through the FIFA rule book, has allowed balls with an outer layer made from polyurethane or polyvinyl chloride, which have been around since the 1960s.[38] The American football helmet is made of molded polycarbonate. Baseball bats, historically made of wood and later aluminum, now have various forms of polypropylene, high-density polyethylene, or low-density polyethylene for youth baseball leagues. Swimming pools are primarily lined with vinyl resin plastic, while pool toys are nearly all made of various plastic resins.

In the great outdoors, athletes enjoy fishing with monofilament fishing lines, and it is common to find recycling locations near docks for fishing lines. (Fishing lines are a widespread plastic marine debris.) Water boots are most likely coated with neoprene, also known as polychloroprene, a synthetic latex rubber resistant to water, oils, and solvents. Boats are primarily made of fiberglass, replacing the hardwoods of the past. Camping gear is now made of flexible and waterproof polyester, vinyl, and polypropylene. The ubiquitous frisbee, first marketed as a plastic flying disc in 1948 and inducted into the National Toy Hall of Fame in 1998,[39] is made of flexible polypropylene plastic.

The list can be exhaustive, covering every building, every room, and everything you do. Consider the plastics in your tools, sewing cabinet, crafts, and hobbies. Consider the plastics in the automobile, the trim around the body and the parts within the car. Look at the plastics around the office—in the elevator, the staircase, the snack room, and the conference room. Plastics abound in the schools— the classroom, the lunchroom, the halls, and the gymnasium. Look around your desk at the writing instruments, the pens, the notebooks, the book sleeves, the file folders and organizers.

A PLASTICS WORLD

Plastic substitutes for most basic materials such as metals, glass, wood, and cloth fibers. When you realize the amount of plastic surrounding you, even on your own body (watches, glasses, clothing, buttons, zippers, shoes), you wonder if there is an invasion of this substance called plastics. It reminds me of the film *The Blob* (directed by Irvin Yeaworth and written by Kay Linaker and Theodore Simonson).[40] The film's plot involves an alien "blob" that invades a small town, slowly seeping into people's homes. The rest of the plot mostly does not apply here; however, consider the metaphor of a plastics alien seeping into our daily lives and slowly replacing our common natural materials of the past, creating a plastics world. That is the world we live in today.

In 2019, an estimated 368 million metric tons of plastic were produced globally. The United States contributes 38 million metric tons of that plastic every year, and Britain contributes an estimated 1.7 million metric tons. We live in a plastics world—a plastics ocean world where the Great Pacific Garbage Patch (one of six) is around 1.6 million square kilometers—larger than the state of Texas. Looking into the distant future, how will a paleontologist view our society's fascination with plastics? "Plastic will definitely be a signature technofossil," says paleontologist Sarah Gabbott, of the University of Leicester. "So wherever those future civilizations dig, they are going to find plastic. There will be a plastic signal that will wrap around the globe."[41]

A friend of mine in the recycling industry, Chaz Miller, was recently quoted defending the plastics industry, stating, "According to Plastics News' most recent ranking of Plastics Recyclers/Brokers, the top ten firms reprocessed 4.27 billion pounds of recycled plastics in 2022. Almost half were postconsumer plastics, the rest postindustrial."[42] If you convert the pounds to tons, that is 2.1 million tons of plastics recycled, compared to an estimated 368 million tons produced annually. Chaz may be referencing a particular segment of the recycling industry. However, several recycling industry experts estimate that less than 9 percent of the labeled plastics on consumer store shelves are recycled in today's recycling markets.[43] I have not seen a higher estimate. The plastics industry will cry foul at that comparison, as I group all plastics produced, not just those that can be recycled (e.g., toothbrushes and kids' toys cannot be recycled). Yet, that is my very point: The plastics recycling industry cannot recycle the vast amounts of plastics being produced.

In 2010, the American Chemistry Council estimated that about 8 percent of global oil and gas supplies were used as feedstock for the production of plastics.[44] A 2018 report from the International Energy Agency estimates that 12 percent of the total demand for oil in 2017 was for the production of plastics, accounting for 12 million barrels per day.[45] The report predicted a 100 percent increase (doubling) over the next thirty years due to the increased use of plastics.

No, I don't subscribe to a plastics conspiracy theory like the alien blob. Still, I wish to call out the nature of our world today: the common everyday reliance on fossil fuel plastics within every facet of our lives. You cannot turn your head in any direction without seeing plastics in some form, no matter where you are. Plastics abound and surround us everywhere in every part of our lives. The point? All these plastics are produced from fossil fuels, most likely not to be reused or recycled, and their end-of-life may be ocean-side, in an environmentally devastating manner. If the solution to climate change is less reliance on fossil fuels, shouldn't we focus on this overreliance on fossil-fuel-generated plastics?

This chapter began with a statement on the stock valuation of Dow Chemical and its involvement in plastics. I offer one additional tidbit of information related to the economic valuation of plastics. To quote energy and environment researcher Ian Tiseo, "The global market value of plastics is forecast to grow from $523 billion U.S. dollars in 2018 to more than $750 billion U.S. dollars in 2027. Plastics are one of the most ubiquitous artificial materials on Earth."[46]

Now imagine how that would measure not in stock profits but in environmental harm.

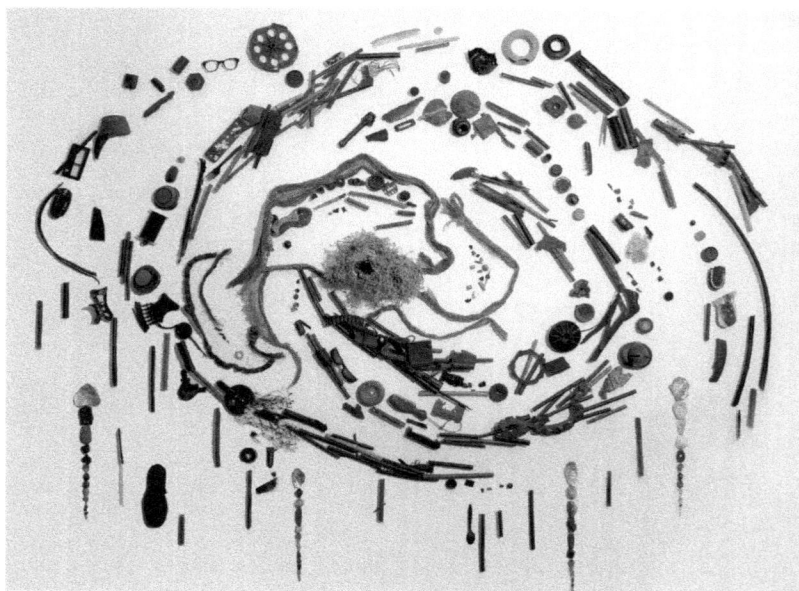

Artwork by Pam Longobardi + Drifters Project

SUBMERGENCE/EMERGENCE, 2015

Drift net, ocean plastic, propeller fouling, minerals, bryozoans, shark bites, specimen pins, silicone, abalone shell, and minerals, 96" x 56" x 14". Special commission memorializing the survival of a family whose boat sunk from a drift net propeller entanglement off Maui. The turquoise net at center is an actual propeller fouling artifact of fused plastic drift net that I found in 2006 on the Big Island Hawaii. Such propeller fouling is a common hazard for boaters, often causing the failure of the engine. The family survived after the captain abandoned the ship but left a survival raft, which they floated in for several hours until they were discovered by Hawaiian fishermen who called the Coast Guard to rescue them. Collection of Sean and Martha Cook.

—PAM LONGOBARDI

Chapter 3

PLASTIC RESINS AND THE PLASTICS CODING SYSTEM

"Nature is with us if we can learn how to align with it
and not break the basic laws that generate life."

—FRANCES MOORE LAPPÉ

My grandfather had an abundant family farm, including several field crops, a large vegetable garden, and several fruit trees. Over the various harvest seasons, I had the privilege to assist my grandmother in canning the tomatoes and the grapes made into jellies. I enjoyed learning how to make applesauce, but tried my best to avoid the peeling chores. Throughout the fruit and vegetable canning work, we used no plastic containers, only glass jars. There was no use of plastic wraps, cling wraps, plastic Ziploc bags, or plastic freezer bags. We utilized paper wraps, steel cans, and glass jars, and stored them in the cold cellar. Today's world has changed from those days—but has it changed for the better?

In the last chapter, various plastic resins were mentioned in their uses for making industrial and consumer products and packaging.

What is plastic resin? How many different types of plastic resins are there? Are products made of plastic resins recyclable, and what does the chasing arrow symbol mean on specific plastic packages and products?

The virgin feedstock material used to manufacture plastic products and packaging is called a resin. It is labeled a *virgin feedstock* because it is freshly manufactured, new, and not recycled or reused in origin. It is labeled a *resin* because it is a composite of plastic molecules and additives. The additives change the characteristics of the plastic polymers by making them harden in a certain way or giving them strength or flexibility, depending on the technical needs of the product's end use.

Plastic hydrocarbons are a product of oil or natural gas refinement. This is an essential point of this book: *Plastics are fossil fuel products.* The crude oil—or natural gas—must be distilled through a heat-intensive process, whereby different chemicals are separated. Some of the resulting liquid is then placed through a process called "cracking," which extracts a polymer hydrocarbon compound. A resin is formed when this polymer compound is mixed with additives and a catalyst. Additives may include color dyes, flame-retardant chemicals, and other chemicals that may offer desired characteristics. These plastic resins can be utilized to make plastic products and packaging through thermoplastic or thermosetting processes with various molding techniques.[1]

To provide an example of the various forms that plastic resins can take, consider the plastic polyethylene. The different classifications of polyethylene, based on weight, density, and other characteristics, offer these separate categories of polyethylene resins:

- Ultra-high-molecular-weight polyethylene (UHMWPE)

- Ultra-low-molecular-weight polyethylene (ULMWPE or PE-WAX)

- High-molecular-weight polyethylene (HMWPE)

- High-density polyethylene (HDPE)

- High-density cross-linked polyethylene (HDXLPE)

- Cross-linked polyethylene (PEX or XLPE)

- Medium-density polyethylene (MDPE)

- Linear low-density polyethylene (LLDPE)

- Low-density polyethylene (LDPE)

- Very-low-density polyethylene (VLDPE)

- Chlorinated polyethylene (CPE)[2]

Although there may be hundreds of arrays of polymer hydrocarbon chains, plastic resins can be aligned into common groupings. The plastics industry saw a need to categorize these resins into common groupings for ease of identification. Thus, a trade association called the Society of the Plastics Industry (SPI) called for a standard listing and coding system.

COMMON PLASTIC RESINS

The following table showing standard plastic resins, listed by resin codes, and the end product use of the plastics was created based on information from the Science History Institute website.[3] Note that the uses listed are not about recycling and recyclability, but about the original products or packaging made from that plastic resin type.

Recycling Codes for Plastic Resins

Recycling code	Polymer and structure	Uses
01 PETE	$-O-CH_2-CH_2-O-C-\langle\bigcirc\rangle-C-$ with O below each C Polyethylene terephthalate (PET) 29% overall recycling rate[4]	Bottles for soft drinks and other beverages
02 HDPE	$-CH_2-CH_2-CH_2-CH_2-$ High-density polyethylene 29% overall recycling rate	Containers for milk and other beverages, squeeze bottles
03 PVC	$-CH_2-CH-CH_2-CH-$ with Cl below each CH Vinyl/polyvinyl chloride <1% overall recycling rate	Bottles for cleaning materials, some shampoo bottles
04 LDPE	$-CH_2-CH_2-CH_2-CH_2-$ Low-density polyethylene <1% overall recycling rate	Plastic bags, some plastic wraps
05 PP	$-CH_2-CH-CH_2-CH-$ with CH_3 below each CH Polypropylene 3% overall recycling rate	Heavy-duty microwavable containers
06 PS	$-CH_2-CH-CH_2-CH-$ with benzene ring below each CH Polystyrene <1% overall recycling rate	Beverage/foam cups, toys, clear envelope windows
07 Other	All other resins, layered multi-materials, and some containers <1% overall recycling rate	Some ketchup bottles, snack packs, and products where the top differs from the bottom

PLASTIC RESIN CODING

So what does the chasing arrow symbol with a number in the center mean on specific plastic packages and products? Prior to plastic product coding, the chasing arrow symbol was first used in the 1920s to inform the public that paper products were recyclable. Over the decades, recycling chasing arrows were utilized to educate the public on the recyclability of a wide range of materials, from paper, newspaper, and cardboard to also include glass bottles, aluminum cans, steel cans—and eventually, plastic bottles.

In 1988, the Society of the Plastics Industry introduced its voluntary resin identification coding system, which included a plastic resin number inside the recycling chasing arrows. Because the recycling industry did not adopt this coding system voluntarily, SPI lobbied throughout the United States for state regulations to adopt these plastic resin codes and require that state- or market-specific plastic containers and packaging use this coded symbol, indicating to consumers that the plastic could be—and would be—recycled. For the most part, SPI succeeded. Yet in most cases, the local recycling systems within the states were not consulted or made aware of the adoption of this new numeric coding system. An NPR news story noted that the plastics industry documents "show that just a couple of years earlier, starting in 1989, oil and plastics executives began a quiet campaign to lobby almost 40 states to mandate that the symbol appears on all plastic—even if there was no way to recycle it economically."[5]

When the coding symbol began to hit the marketplace in the early 1990s, Coy Smith operated a recycling collection facility near San Diego. He was quoted in the same NPR news story noted previously, stating, "The consumers were confused. It undermined our credibility, undermined what we knew was the truth in our community, not the truth from a lobbying group out of D.C."[6]

The Society of the Plastics Industry received a report in 1993 that told them about the problems in the recycling field. As reported by NPR, the report said bluntly, "The code is being misused. Companies are using it as a 'green' marketing tool."[7] The report told them that the resin code created "unrealistic expectations" about how much plastic could be recycled. Despite nationwide objections from recycling facilities, the coding system remained unchanged.

The American Chemistry Council's website notes, "The numbers and letters are intended as resin identification codes to facilitate the recycling process."[8] However, the various recycling material recovery facilities that received and sorted plastics continued to describe their acceptable recyclables through product descriptions and pictorials, avoiding using the numerical symbols. Local governmental agencies in charge of recycling education received thousands of calls, so they eventually had to address the resin coding system and translate the regional program's acceptable list through the coding system. The gap was bridged reluctantly.

Over time, manufacturers of plastic products began to embed the plastic coding symbol on nearly all plastic products and packaging, far beyond the intended reach of the initial coding system. Since no legal or scientific enforcement agency was declared to deal with false recycling claims, there became no means to determine if the product or packaging with the symbol was recyclable. Some recycling industry experts estimate that less than 9 percent of the labeled plastics on consumer store shelves are actually recycled in today's recycling markets.[9]

To propel this new plastic coding system, SPI provided sample legislative language, which was adopted in full text by many states, supported by testimony that this legislation would dramatically increase plastic recycling rates. A total of thirty-nine states have enacted the SPI plastic container coding requirements. Here is an example of a state plastic resin coding law, from the State of Ohio:

Ohio Revised Code/Title 37 Health-Safety-Morals/Chapter 3734 Solid and Hazardous Wastes

Section 3734.60 | Plastic containers labeled with code for basic material used in bottle or container.

Effective: November 2, 1989

(A) As used in this section:

(1) "Label" means a molded imprint or raised symbol that includes a code consisting of letters and numbers and is placed on or near the bottom of a plastic bottle or rigid plastic container to indicate the plastic resin used to produce the bottle or container.

(2) "Plastic" means any material made of polymeric organic compounds and additives that can be shaped by means of the flowing of the material.

(3) "Plastic bottle" means a plastic container that has a neck that is smaller than the body of the container; that accepts a screw-type cap, snap cap, or other closure; and that has a capacity of at least sixteen ounces but less than five gallons.

(4) "Rigid plastic container" means any formed or molded container, other than a plastic bottle, that is intended for a single use, is composed primarily of plastic resin, has a relatively inflexible finite shape or form, and has a capacity of at least eight ounces but less than five gallons.

(B) On and after January 1, 1991, no person shall either manufacture or distribute for use in this state any new, unfilled plastic bottle or rigid plastic container unless it bears a label with the appropriate code as prescribed in this section that indicates the plastic resin used to produce the bottle or container. A plastic bottle or rigid plastic container having a label or basecup composed of material different from that comprising the rest of the bottle or container shall be labeled with the code for the basic material used in the bottle or container.

The label required by this section shall consist of an equilateral triangle formed by three curved arrows of short radius with the apex of each point of the triangle at the midpoint of each arrow. The head of each arrow shall be at the midpoint of each side of the triangle with a short gap between the head of the arrow and the base of the succeeding arrow. The triangle formed by the three curved arrows shall depict a clockwise path around the code number, which shall be placed at the center of the triangle. The code letters shall be placed immediately below the triangle. The following code numbers and letters shall be used on the labels:

(1) For polyethylene terephthalate, the letters "PETE" and the number "1";

(2) For high density polyethylene, the letters "HDPE" and the number "2";

(3) For vinyl, the letter "V" and the number "3";

(4) For low density polyethylene, the letters "LDPE" and the number "4";

(5) For polypropylene, the letters "PP" and the number "5";

(6) For polystyrene, the letters "PS" and the number "6";

(7) For any plastic material named in rules adopted under division (C) of this section, the code letter and code number prescribed for the plastic material in those rules;

(8) For any other plastic, including, without limitation, multi-layer materials, the word "OTHER" and the number "7".

(C) When the director of environmental protection considers it appropriate, he may adopt rules in accordance with Chapter 119. of the Revised Code listing plastic materials in addition to those listed in divisions (B)(1) to (6) of this section and prescribing a code letter and code number for each of those additional plastic materials. When labeling requirements similar to those established by this section have been established pursuant to the laws of other states or any such additional material, the code number and code letter for that material established under this division shall be consistent with the code number and code letter for that material established pursuant to the laws of those other states.

(D) The environmental protection agency shall maintain a list of the codes prescribed in divisions (B)(1) to (8) of this section and shall provide a copy of the list to any person upon request.[10]

In addition to the state code requiring the plastic container coding, the Advancing Standards Transforming Markets (ASTM) offer manufacturers advice on how to interpret these plastic codes. Here is an example from ASTM's website:

Resin Identification Codes are not "recycle codes." The Resin Identification Code is, though, an aid to recycling. The use of a Resin Identification Code on a manufactured plastic article does not imply that the article is recycled or that there are systems in place to effectively process the article for reclamation or re-use. The term "recyclable" or other environmental claims shall not be placed in proximity to the Code.[11]

As noted previously, recyclers disagree with this characterization of the plastic coding system, as recycling materials recovery facility (MRF) operators generally do not utilize the coding system. This type of public education has encouraged "wish-recycling," where consumers add nonrecyclable plastics to their bins, wishing that the recycling program would recycle them.

Nearly all thirty-nine states that have adopted the SPI plastic container codes have adopted this language, with few variations, proving the effectiveness of SPI's lobbying power. For the most part, recycling MRF operators only became aware of this legislative requirement after the residential recyclers in their area started delivering the coded plastic containers, even though most *are not recyclable by current recycling markets*! The critical point here is that plastics are recyclable only if there is a market to which the MRF operator can ship. There, the recycled material can be utilized as a feedstock to make a clean, needed product or packaging material. If that market does not exist, that plastic product is not recyclable, even though it may be "recyclable" in a technical sense.

Another critical point on recycling markets before we move on: Who pays for recycling the plastic containers received at the recycling MRF? Residents and local governments pay for recycling collection and processing; thus, this coding system victimizes utility rate payers. The SPI coded the containers without consulting recycling MRF operators or doing any research on the coding system's effects on the plastics end-markets. There is also concern that local governments need to be in the loop. In short, the coding system was developed in-house rather than through collaboration with the entire recycling supply chain. For the American Chemistry Council (ACC) and the SPI to claim that this coding system "facilitates the recycling process" is beyond the belief and understanding of the professional recycling community.

The thirty-nine U.S. states that adopted the coding system into regulation represent a large enough economic market force that the entire plastics product and packaging industry began imprinting the codes onto products for distribution all across the United States. It would be impractical to segregate shipments to non-coded states. Beyond the initial concerns about the labeling process, the plastic container and packaging industries have learned that no professional, governmental oversight exists to enforce these code regulations. Without oversight, there was no guidance on which containers required the codes and which new products and packaging did not. The plastics manufacturers played the safe bet and began marking nearly all plastic products and packaging with the coding. Some may claim this was an effort at "greenwashing" plastics as environmentally friendly.

With the explosion of products utilizing the plastics coding, the new recycling advertising from SPI pushed the codes to identify what is recyclable. Recycling collection programs began to be overwhelmed with plastic, even if the local program could not recycle the incoming plastics. An example was a program that could recycle PET plastic bottles but needed a market for non-container PET. As PET is labeled #1 in the recycling chasing arrows, consumers started delivering all #1 PET products and packaging to that recycling center, even though they may not be recyclable PET containers. This forces the recycling program to invest in new recycling sortation equipment (or additional staff) to separate the marketable material from the non-marketable plastics. The marketable PET gets baled and shipped to the market. The non-marketable plastics are then sent to a disposal facility. This all adds up to more costs, less efficient recycling, and more landfilling.

How did the recycling industry work this plastic flow out? Unfortunately, not by eliminating the SPI codes. Instead, an unlikely solution arose. In the early 2000s, China's economic growth was stunning and matched by a strong need for recyclables to build its new economy. China plastic buyers popped up ready and willing to buy

American baled plastics. Later, we found out about the environmental hazards piling up in China and Southeast Asia. This was not an intelligent plan to find recycling markets. Still, little was outwardly known about the plastics end markets throughout the 1990s—the plastic manufacturers were just pleased to have found an end market for the various SPI-coded plastics flooding their facilities. (More on the China markets in a later chapter.)

HOW DO WE "FIX" THE PLASTIC CODING SYSTEM?

Due to various factors, including Americans' "wish-recycling" because of the confusion generated by the plastic coding system and China's "send-me-all" market that gave recycling MRF operators an easy out to collect and ship all plastics to China, big issues arose when the shipment pipeline shut off abruptly in 2020. We now have a plastics problem, largely unreported, of a vastly higher production curve of new plastics on the markets and higher disposal rates. Let's stop pretending—these plastics are not being recycled.

As *Grist* first reported in an interview, Jennie Romer, then the Environmental Protection Agency's deputy assistant administrator for pollution prevention, called for the FTC to end the "deceptive" use of the chasing arrows on plastics. "There's a big opportunity for the Federal Trade Commission to make those updates [to its Green Guide] to really set a high bar for what can be marketed as recyclable," Romer said. "Because that symbol, or marketing something as recyclable, is very valuable."[12]

Chemical engineer Jan Dell pushed a shareholder proposal against Kraft Heinz for printing the chasing arrow on its packaging even though it was not recyclable. Kraft Heinz is "greenwashing their products to be recyclable when they're really not," Dell said in an interview with *Inside Climate News*. "I'm holding them to truth and accountability. As a shareholder, I'm worried the company's brand value, which directly affects sales and the stock price, could be hit hard by lawsuits on false advertising."[13]

With increasing plastics production and reduced recycling capacity, we're facing a growing problem with no silver bullet solution. We can begin with eliminating the consumer source of confusion—there needs to be clear and concise public education about what is recyclable and what is not. I call for the immediate elimination of the plastic coding system. This is not an easy task, for it takes the education of forty-five state legislative bodies, an increase from the original thirty-nine. But the truth must be told that this coding system is causing much harm to the American recycling system. The removal of the existing system is an important first step, and no replacement coding system is needed.

The forty-five states that have embedded these plastic resin codes in their state legislative codes are AK, AR, AZ, CO, DE, FL, GA, HI, IA, IL, IN, KS, KY, LA, MA, MD, ME, MI, MI, MN, MO, MS, NC, NE, NJ, NV, OH, OK, RI, SC, SD, TN, TX, WA, and WI.[14] The State of California set a great example by passing the Truth in Recycling Law, which eliminates the SPI plastic coding requirement and establishes a set of requirements for manufacturers to provide proof their package or product is recyclable before they can utilize the recycling symbol in the state.[15] This legislation can be a model for citizen action in the states that still require use of the SPI resin codes.

The confusion over plastics is not the consumer's fault. The system is rigged to frustrate you, so you take one of two frustrated actions: throw everything in the recycling bin or give up on recycling. I ask you to take a different path. Challenge the plastics industry. Challenge the plastics coding system. The recycling world is better off without plastic codes.

Artwork by Pam Longobardi + Drifters Project

CRIME OF WILLFUL NEGLECT, 2014

Four hundred twenty-nine pieces of vagrant oceanic plastic from Greece, Hawaii, Costa Rica, and the Gulf of Mexico, 84" x 138" x 6". Black plastic is the most ubiquitous and least recyclable type of plastic and most visibly speaks of plastic's dark nature by exhibiting its origins in oil and fossil fuels. Petroleum companies go to more and more extreme lengths to extract the last remaining drops of oil from miles underneath the ocean and under former Arctic ice while the world ocean is plagued by discarded plastic. Plastic objects are the cultural archeology of our time, a future storehouse of oil, and the future fossils of the Anthropocene. The Deepwater Horizon Disaster is a crime that has not seen full justice and whose future long-term damage to the Gulf of Mexico continues to unfold.

—PAM LONGOBARDI

Chapter 4

PLASTICS IMPACTS ON HUMANS AND THE ENVIRONMENT

"I'm not a teacher, but an awakener."

—ROBERT FROST

As a teenager, I would often walk the family farm—sometimes in deep conversation with my grandfather, sometimes just walking alone, musing in my adolescent world. The typical walk included picking up trash that had accumulated on the edges of the farm, usually litter thrown from passing cars. From years of observation, the most typical litter I picked up was fast-food debris, and the second most common item was plastic shopping bags.

Plastic single-use bags were first introduced in the local gas station convenience stores before the groceries nearby converted their bags from paper to plastic. Almost immediately, we saw them in the litter streams on the farms. If the plows or harvesters were to pick up the bags, they would become entangled in the gears and cause damage, so it became essential to pick up the litter. It used to be that the most critical danger was glass bottles, but soon it became the menace of plastic bags.

Plastics have invaded our environment in many ways over so many years—are we just now *awakening* to this damage?

Americans produce, consume, and dispose of more plastic than any other nation in the world. Is that what it means to be "number one"? Based on a research study published in 2020, "In 2016, the U.S. population produced the largest mass of plastic waste of any country in the world and also had the largest annual plastic waste generation of the top plastic waste-generating countries [42.0 million metric tons]."[1]

According to the U.S. Environmental Protection Agency (USEPA), in 2018 (the latest statistics available), plastics disposal represented 12.2 percent of municipal landfill volume. For the United States, plastic disposal happened through landfills (75.4 percent), incineration (15.3 percent), and recycling (9.3 percent).[2] Unreported in these USEPA stats was plastic litter.

The "leakage" of plastic volume into our environment, streams, lakes, and oceans is often not measured. The academic study mentioned previously attempted to statistically measure "mismanaged plastic waste" within the United States that was not properly collected and estimated the 2016 volume to be "between 0.98 and 1.26 million metric tons."[3] In more familiar terms, that is approximately 1,260,000,000 metric tons or 2,772,000,000,000 pounds of plastic waste, or the equivalent of 55,440,000,000,000 single-serving 500 ml empty plastic water bottles. Mind-boggling.

National Geographic, commenting on another report that came to the same conclusive measure of 42.0 million metric tons of plastic waste generation, added, "The U.S. also ranks as high as third among coastal nations for contributing litter, illegally dumped trash and other mismanaged waste to its shorelines."[4] This litter and illegally dumped trash (including plastics) find their way to our lakes, streams, and oceans, affecting aquatic life, drinking water, and the quality of the fish we eat.

Plastics is a fossil fuel product, with its production beginning at the wellhead, coal mine, or drill pads. From cradle to grave, plastics have environmental and health impacts. In a larger sense, there is no actual "grave" for plastics, as the entire spectrum of plastics never totally decomposes but instead breaks into smaller and smaller pieces, having a considerable impact on the air, water, soils, and even our human bodies. The large volume of plastic "leakage" into our environment significantly impacts all life on this planet.

THE IMPACTS OF PLASTICS ON HUMANS

The primary sources of plastic production in the United States are the fracking of natural gas or the refinement of oil. In 2020, more than 500,000 fracking wells were drilled in the United States, according to the USEPA. Hydraulic fracking wells utilize large quantities of water and chemicals that are sandblasted into impermeable rock formations at pressures high enough to crack the rock, allowing the trapped gas and oil to flow to the surface. The USEPA has identified more than one thousand chemicals utilized in the various hydraulic fracking wells throughout the United States, many of which are hazardous to humans and the environment.[5]

From that same report and additional studies in the following years, the USEPA found evidence of groundwater impacts. Surface chemical spills reaching groundwater and local drinking water resources, unmitigated releases of gases and liquids into the environment, and disposal of hydraulic fracturing wastewater in unlined pits resulted in groundwater contamination.[6] This mismanagement of hazardous chemicals offers direct and harmful exposure to workers and those in the outlying communities.

Fracking communities often report smog created by the air emissions from the hydraulic fracturing, which releases toxic air contaminants such as benzene, toluene, ethylbenzene, and xylene; fine particulate matter (PM2.5); hydrogen sulfide; silica dust; and nitrogen oxides and volatile organic compounds.[7] These reports note the human health effects of air emissions, including respiratory and neurological problems, cardiovascular damage, endocrine disruption, congenital disabilities, cancer, and premature mortality. Gas and oil industry workers face the risks from on-site exposure to higher concentrations of toxic chemicals and other airborne materials, including silica, which can lead to lung disease and cancer when inhaled.[8] Additional human health exposures result from fracking waste containing salts, chemicals, and radionuclides, often spread on roads or dumped into drainage areas for ease of disposal.

Communities living near fracking facilities have been studied for health risks due to the USEPA reports of air, water, and soil contaminants. Environmental health experts note that these studies have found "that living near fracking wells increases the risk of premature births, high-risk pregnancies, asthma, migraines, fatigue, nasal and sinus symptoms, skin disorders, and heart failure; and laboratory studies have linked chemicals used in the fracking fluid to endocrine disruption—which can cause hormone imbalance, reproductive harm, early puberty, brain and behavior problems, improper immune function, and cancer."[9]

Fracking is just the start of the harmful, potentially hazardous processes involved in producing plastics. Gas and oil are then transported via pipeline, trucks, and rail from the fracking and oil refineries to the production facilities. Then, superheated water is used to "crack" the methane (CH_4) molecule into ethane, ethylene, propane, and other products used to make a variety of polymers. Additives are mixed in to form the plastic resins, again involving chemical substances that are hazardous to humans and the local environment. There are significant claims that the petrochemical additives (bisphenol A, lead, brominated flame retardants) "can cause nervous system disorders, reproductive impairments, developmental problems, cancer, and genetic impacts. These conditions result from inhalation, ingestion, and dermal absorption."[10] Plastic production often utilizes toxic chemicals such as 1,3-butadiene, benzene, styrene, toluene, ethane, propylene, and propylene oxide. Again, these chemicals "are often carcinogenic and found in higher concentrations along 'fence line' communities adjacent to industrial sites."[11]

Once the plastic resins are produced, they are shaped into plastic flakes, pellets, and nurdles—today's most common trading form of feedstock plastic in the U.S. market. Both recycled content and virgin petrochemical forms (plastic resins) are shaped this way to serve as direct feedstock in most plastic injections and pressure-forming facilities.

At these plastic resin production facilities, human contact with the material is inevitable. Again, the areas of concern lie in hazardous chemical spills and air emissions from petrol-chemicals. In 1990, the USEPA reported to Congress that it "found spilled pellets had become 'ubiquitous' in the environment." The report blamed the "oil and gas industry, pellet manufacturers, and transporters as possible culprits in the spills."[12] These spilled plastic pellets impact the natural wildlife in the area of these plants, but as the supervisors order cleanups, the workers come in direct contact with the plastic pellets too. In 2020, following these pellet spills, NPR noted, "New research suggests more than 230,000 tons of [plastic] pellets enter the ocean each year, contaminating the water and sickening birds, fish, and other wildlife."[13]

In September 2024, a railcar in Addyston, Ohio, a western suburb of Cincinnati, triggered a release valve of styrene due to excessive heat. During an unusual local heat wave, the tanker railcar held styrene for an unknown period, perhaps a week. The triggered valve released styrene gas into the surrounding neighborhoods for several hours as firefighters pointed water hoses on the tanker to cool it down. Hundreds of residents, concerned about gas inhalation risks, were evaluated at community centers.

Styrene is a compound often used as a precursor for making polystyrene and numerous other copolymers. As it is categorized as part of the vinyl group, it can polymerize to create various different materials, such as insulation, carpet backing, fiberglass, pipes, automobile and boat parts, and plastic food containers.[14] It evaporates readily, has a "sweet" smell, and is liquid at ambient temperature but volatile and explosive at high temperatures.

The local newspaper reported, "Styrene, which health officials say is likely to cause cancer in humans, was spewing out from the rail car over four miles from its destination: a plant owned by the chemical company Ineos Styrolution in Addyston."[15] Amy Townsend-Small (a University of Cincinnati professor quoted for the news story) believed this likely wouldn't be the last time it happened and expected the amount of toxic chemicals transported via rail to increase "because of what she believes is a lack of federal oversight on fracking."[16]

PFOA AND PFAS IMPACTING HUMAN HEALTH

Moving from the production facilities to the consumer level, consider the impacts of cooking with plastics. Consumers who use plastics for food preparation or put their drinks in plastic cups are exposed to toxic plasticizers, which can leach out of the plastic into our foods when plastics are exposed to heat and acidic or alkaline foods.[17]

An entry-level impact on our bodies is through the cooking instruments we use on the stovetop. In 2000, Rob Bilott, a partner at the law firm Taft Stettinius & Hollister, LLP, discovered the role of PFOA, a polymer or a form of plastic, in the "poisoning of the drinking water of 70,000 residents" in a small town in West Virginia. The case is described in his book *Exposure: Poisoned Water, Corporate Greed, and*

One Lawyer's Twenty-Year Battle against DuPont.[18] Through the discovery phase of the legal trials, it was revealed that PFOAs were utilized in the manufacturing of consumer products, including as a coating on frying pans to make them nonstick, under the patented trade name of Teflon. Although not used in the same sense as a formed plastic straw or plastic cup, instead, PFOA is used as surface additive or coating in consumer and industrial products, and the desired effect is to repel water or oil. Common uses beyond the frying pan include carpets, raincoats, and boots. The lawsuit displayed that a community's drinking water was poisoned with PFOAs, connected to the company's manufacturing processes, with linkages to a variety of cancers experienced in the town.

The health concern with PFOA and related plastic chemicals was the eventual possible ingestion of residual traces of the chemicals. Exposure has been linked to thyroid disorders, chronic kidney disease, liver disease, and testicular cancer.[19] PFOA is also found in the blood of more than 98 percent of people who took part in the U.S. 1999–2000 National Health and Nutrition Examination Survey.[20] The USEPA phased the product out by 2015, although it is still commonly used in products manufactured in other countries.[21]

Even with these phase-outs, the damage has already been done. PFAS (poly-fluoroalkyl substances) and PFOA (perfluorooctanesulfonic acid) are known as "forever chemicals" due to their indestructible and bioaccumulating persistent nature. They will forever represent the more significant categories of petrol-based chemicals. Both have been found in the bloodstream of most Americans in various health studies, as well as a majority of the drinking water systems in the United States, with the persistence of the chemicals indicating that there are no known remedies for removal.[22]

Most plastic products and packaging containing PFAS will most likely be disposed of in a landfill at the end of their life. Landfill operators suggest that their landfills function as "sequestration vehicles" for

PFAS. A recent scientific study determined the pathways the PFAS traveled upon disposal in a landfill. The researchers tested the liquid leachate and the gas released from three municipal landfills in Florida. Their study findings determined "that landfill gas, a less scrutinized byproduct, serves as a major pathway for the mobility of PFAS from landfills."[23] Notably, a typical landfill captures only 60 percent of released gases, with the remaining unmitigated gas released into the local residential neighborhoods.

Local governments hear the voices of citizens' concerns regarding the health impacts of plastic packaging. Governor Cuomo of New York signed state legislation banning PFAS from a specific range of food packaging. At the press conference announcing the signing of the state bill, State Assemblywoman Patricia Fahy of Albany sponsored the legislation along with State Senator Brad Hoylman, D-Manhattan. "PFAS—a dangerous and cancer-causing class of chemicals commonly used in everyday food packaging—however, is anything but safe for New Yorkers."[24] That same New York bill also bans the incineration of "Aqueous Film Forming Foam" used in firefighting and containing PFAS chemicals, which raises the question of firefighter exposure to PFAS chemicals when utilizing the foam to fight a fire.

Other local governments are exploring banning PFAS; according to the Safer States bill tracker, twenty-eight U.S. states have proposed policies, and fifteen U.S. states have adopted policies for regulating PFAS as of March 2021.[25] The European Union (EU) set drinking water standards for all PFAS compounds throughout all twenty-seven member nations. Sweden, the Netherlands, Germany, and Denmark are restricting all PFAS compounds under Europe's chemical regulations framework, with phase-out plans by 2030. Denmark has banned all PFAS from food packaging.[26]

In 2024, the USEPA established new rules for the restrictions and limits of PFAS in drinking water. First, the USEPA finalized a national environmental rule to designate two common PFAS—PFOA and PFOS—as hazardous substances under the Comprehensive Environmental Response, Compensation, and Liability Act (CERCLA). Then, the USEPA issued the first-ever national "legally enforceable drinking water standard to protect communities from exposure to harmful PFAS." To assist in implementing these requirements, the USEPA announced a grant fund of $1 billion to help states, territories, and private well owners with PFAS testing and treatment.[27]

Unfortunately, this regulatory structure does not restrict the use of PFAS in consumer products. However, insurance liability for these products is increasing as the regulations become more burdensome and the scientific body of evidence increases. Designating PFAS-related chemicals as hazardous chemicals will assist in restricting use but not eliminating it. In January 2024, the EPA automatically added seven PFAS to the list of chemicals covered by the Toxics Release Inventory (TRI).[28]

CHEMICAL ADDITIVES TO PLASTICS IMPACTING HUMAN HEALTH

Consumers have also grown concerned about degrading plastics in home-based heat sources such as microwaves and dishwashers. Frederic vom Saal, Ph.D., Curators' Distinguished Professor Emeritus of Biological Sciences at the University of Missouri, says, "There is no such thing as safe microwaveable plastic. As you heat it in a microwave oven, the chemical bond degrades. You cannot see this happening. You cannot taste or smell it, but you are getting dosed at a higher and higher amount."[29] Dr. Saal was referring to the chemical known as BPA.

Bisphenol A (BPA) has been a chemical additive in polycarbonic plastics and epoxy resins since the 1960s. Polycarbonate plastic containers store food, beverages, and water bottles. Dr. Saal's research "concerns the effects on fetal development of endogenous sex hormones, naturally occurring estrogenic chemicals in food (such as phytoestrogens in soy), and artificial estrogenic substances in consumer products (such as bisphenol A in plastic)."[30]

Additional research studies detected BPA leaching from plastics into food from microwave heating, while other studies demonstrate the leaching of BPA from dishwashers.[31] Much scientific interest in BPA contamination of food and drink is related to published research suggesting that BPA may be an endocrine-disrupting chemical.[32] Endocrine disruptors act by interfering with the metabolism of naturally occurring hormones. Other research studies have shown that BPA can seep into food or beverages from containers *without heat*. Health professionals have expressed concern about the "possible health effects of BPA on the brain and prostate gland of fetuses, infants, and children."[33]

Many environmentalists consider BPAs to be on the same environmental pathway as the now-banned chemicals DDT and PCBs. A point to note is that plastic containers are the initial carrier of the chemical of concern. If not for using plastic containers for food and beverages, the chemicals would not be carried to the human body. Thus, plastic containers are the root cause of this poisoning.

The Environmental Defense Fund (EDF) has studied the effects of certain toxic chemical additives in plastic food packaging that impact our food. Their research moves beyond the traditional research on BPAs and focuses on plastic packaging additives such as Ortho-phthalates, Perchlorate, Benzophenone and Bisphenol A, B, F, S.[34] The EDF study also notes that they found several heavy metals in the contaminated food samples, possibly sourced from plastic food packaging. I praise the final recommendation of the EDF researchers as they conclude their study with the thought that food packaging manufacturers must "start clean" and be committed to setting "tight virgin-material standards to prevent problematic contamination."[35]

OTHER TOXIC CHEMICAL IMPACTS ON HUMANS

Black plastics recovered from recycling electronic products often contain toxic elements such as antimony, bromine, cadmium, mercury, and lead. The recycled black plastic is remanufactured into consumer toys, clothing, sports equipment, single-use packaging, and trays for food. A scientific study demonstrated that the black plastics from e-waste creates "significant and widespread contamination of black plastic consumer goods ranging from thermos cups and cutlery to tool handles and grips, and from toys and games to spectacle frames and jewelry. The environmental impacts and human exposure routes arising from WEEE [electronic-waste] plastic recycling and contamination of consumer goods . . . including those associated with marine pollution" are significant.[36] The researchers found bromine in approximately half of the black plastics and antimony in a quarter of the samples tested. Lead was the most common contaminant detected in clothing and toys made from recycled black plastics.[37]

Recycled black plastics are commonly used for kitchen utensils and toys, and many contain decaBDE, a banned toxic flame retardant found in household electronics. Lead study author Megan Liu has explored the exposure to contaminated black plastic kitchen utensils, estimating "an average of 34.7 parts per million of decaBDE each day."[38] A kid's toy measured "22,800 parts per million of total flame retardants—almost 3% by weight."[39] The CNN report notes that decaBDE was banned in 2021 by the U.S. Environmental Protection Agency after being linked to cancer, endocrine and thyroid issues, fetal and child development and neurobehavioral function, and reproductive and immune system toxicity.[40]

As black plastics have been proven harmful to humans, what about other intensely pigmented plastics? Researchers from the University of Leicester, in a study focused on how pigmented plastic impacts the environment, have demonstrated that certain deeply colored plastics break down faster, causing concerns for faster plastic degradation into microplastics. "The findings across both studies showed that black, white and silver plastics were largely unaffected whereas blue, green and red samples became very brittle and fragmented over the same time period." The researchers also noted that over the three years of the study, "no brightly colored plastic items were found. But the [beach] sand itself was full of many colored microplastics."[41] Further studies are needed to understand how the color pigments affect food consumption and the human body.

Although it's well known that eating seafood is a healthier diet choice than eating land-based animals, the tides are changing as aquatic life ingest "up to 11,000 tiny pieces of plastic every year," according to scientists at Ghent University in Belgium.[42] In a study sponsored by Plymouth University, about one-third of the fish caught were reported to contain plastic fragments. The European Food Safety Authority cited increased concern for food safety and human health "given the potential for microplastic pollution in edible tissues of commercial fish."[43] As scientific studies continue to document the ingestion of plastics in the various links in our food chains, it is apparent that forms of plastics are becoming a regular daily serving on our dinner plates. That leads to the question: How do these residual micro-plastic particles affect human life?

For decades environmentalists have targeted to-go coffee cups as an unnecessary, unrecycled disposable, and often beneath the surface is a chemical liner in contact with the hot coffee. German journalist Heike Dittmers reported on the inner plastic coating of hot coffee to-go cups. Dittmers collected samples from several fast-serving coffeehouses and took them to a lab for testing. The inner lining, directly in contact with the coffee, contained the "plasticizer diisodecyl phthalate (DIDP, CAS 26761-40-0) chemicals from the coating that can migrate into hot (and particularly into fatty) drinks such as coffee (with milk)."[44] These phthalates are suspected to have endocrine-disrupting effects on humans.

Studies offered by Karolína Brabcová, an expert on plastics and toxic chemicals, "have shown the effect of toxic additives in plastics in the brain development and hormone production of the fetus."[45] She notes in her presentation that 90 percent of all toys made for children worldwide are made from plastic. A 2016 study by the European Environmental Bureau researched the toxicity of 248 toys marketed in Europe. Ninety-two percent were categorized as presenting serious risks of toxic contamination, and 51 percent had elevated levels of phthalates.[46] How many times have we all seen a plastic toy in the mouth of a small child? It is common for a child to chew on plastic toys as they grow new teeth. What impacts does this have on a child's health and development, given the well-documented toxic additives in those plastic toys? It's easy to see the danger here when we consider that the primary base of that plastic toy is a fossil fuel product with chemical additives.

In a study published in the *Journal of the Endocrine Society*, New York University researchers found that in 2018, "the hormone-disrupting effects of plastics in the nation's food and water led to a quarter of a trillion dollars in additional health care costs."[47] The study referenced found that "chemicals can leach directly into food," and as they often "interfere with the body's ability to produce and process fats, phthalates contribute to obesity and diabetes."[48] The study's authors note that their human health estimates are conservative underestimates.

Dr. Leonardo Trasande of New York University has also worked on the human impacts of the exposure to certain plastics and has linked "chemicals . . . used in some plastics to a range of health effects, including developmental problems in children, obesity, and diabetes." Dr. Trasande notes in an interview, "We have strong evidence for human health effects across a small subset of these chemicals, which unfortunately tells us there might be a bigger problem than what we even know now."[49]

I should note, I do not pretend to be a chemical scientist myself. Still, the scientific work of these experts is irrefutable: There is an obvious connection between plastics and human health concerns. The chemical additives to plastic resins that make our consumer products perform to our expectations are slowly harming our health. Science has a slow pathway from hypothesis to study to direct linkage to human health issues, yet an amateur sleuth like me can find hundreds of professional studies providing these connections. And even if the exact cause and effects are difficult to pinpoint, the overall picture of harm is clear. Isn't it time we draw those conclusions and listen to the human cry for change?

PLASTICS IMPACTING THE NATURAL ENVIRONMENT

Both humans and animals have been known to ingest various plastics found in their environment. In India, cows have died from starvation due to their stomachs being clogged with plastic bags they mistook for food.[50] *National Geographic* photographers have captured images of many animals in the wild entangled in plastic waste. In Ethiopia, hyenas listen for garbage trucks at a landfill for their next meal, scavenging through the plastics for edible scraps and sometimes confusing the two. In Japan, hermit crabs are trapped in plastic trash on the beach.

Iconic photos of a plastic straw stuck up a sea turtle's nostril, a dead albatross with its stomach bursting with plastic bags, and a turtle stuck in a six-pack ring are utilized by many environmental groups to plead their case to limit or eliminate single-use plastics in their communities.

The plastics industry has responded by claiming that these are unique situations that are not prevalent, pointing out that the same few photos are used throughout the environmental community. However, numerous different photos are discoverable daily if you search for them. And the visual images are alarming. Wild animals are being mangled, to a great degree, by discarded plastics in our environment. The damage is worldwide throughout the animal kingdom.

Often when hiking in Yosemite and other national parks, I have encountered various single-use plastic bottles thrown off the pathway by careless tourist hikers. Unfortunately, it is not one or two, but many hundreds that accumulate along those well-worn hiking pathways. The park rangers have many stories and photos of natural wildlife being injured by these plastic bottles and packaging as they are mistaken for safe food and drink. Another hazard to ponder, though it's not often reported in the news, is the massive amounts of groundwater consumed by the companies that bottle these single-use plastic containers with water from within our park systems.

In 2018, Nestlé gained an increase in their groundwater withdrawal from 250 gallons per minute to 400 gallons per minute for its White Pine Springs well from a State of Michigan permit. Through the public permitting process, 81,862 comments were logged against the permit, and 75 comments were in favor. Yet the state approved the increased use of local groundwater despite local opposition.[51] In Florida, Nestlé and Seven Springs Water requested a state permit renewal to pull a million gallons of water a day from springs that draw water from the Santa Fe River. Despite resident objections, an administrative law judge from the Suwannee River Water Management District ruled in favor of the permit applicants.[52] Beyond a permit application fee, Florida does not charge water companies for drawing water from the local springs.

Across the pond, litter collection crews in southern England were collecting strange "stones" off the beaches. Upon further inspection, they discovered that the "stones" were made of two forms of plastic: polyethylene and polypropylene. They were composites mixed with chemical additives and heavy metals such as lead and chromium. The researchers studying these plastic stones have called them "pyroplastics," a new form of plastic pollution that was transformed by fire.[53] The precise origin is still unknown, but similar plastic stones labeled "plastiglomerates" were discovered on a beach in Hawaii about a decade ago.[54] These stones appeared to be ocean-littered plastics chemically bound to rock through sunlight irradiation that oxidized the plastic. Both cases present troubling implications. Some of the samples studied from England indicate that beach animals may have ingested these plastic stones, suggesting the introduction of heavy metals into the food chain. And scientists were alarmed that the formation of plastiglomerates could decrease microbial diversity and increase the dispersion of microplastics. "Plastiglomerate poses an imminent danger to ocean sustainability, blue economy, and overall human health," one scientist told *Newsweek*.[55]

PLASTICS IMPACTING THE OCEANS

Now let us move from land to sea. The oceans account for 70 percent of the global surface area of Earth, and 78 percent of Earth's animal biomass lives in the marine environment. The oceans contain 97 percent of the world's water and significantly regulate Earth's climate. Recent scientific studies indicate that 17.6 billion pounds of plastic are dumped into the oceans annually from land-based sources. The visual equivalent of that would be a trash truck full of plastic being dumped every minute of every hour of every day.[56]

Many readers will have already heard of the Great Pacific Garbage Patch. Several recent studies have noted that this "gyre patch" is as large as 1.6 million square kilometers and floats in the Pacific Ocean between Hawaii and California.[57] It consists of all types of garbage but mainly of plastic particles, some floating and some entangled below the surface, as if constructed as a semi-solid, soupy iceberg. Some marine travelers say it's such a thick and substantial surface that you can walk on it for meters. However, the outer perimeters are mostly loose floating bottle caps, plastic ropes, and fishing nets recently "acquired" by the gyre as it grows and captures more material.

It is important to note that ocean gyres naturally circulate ocean currents through the force of nature. There are five ocean gyres: the North Pacific Gyre, the South Pacific Gyre, the North Atlantic Gyre, the South Atlantic Gyre, and the Indian Ocean Gyre. These gyres are commonly referenced as "oceanic conveyor belts" that move and circulate the ocean waters in each of their native areas of the globe. What is *not natural* is the trash these gyres are currently carrying.

The headliner is the North Pacific Gyre, which popular news has named the Great Pacific Garbage Patch. However, two additional garbage patches are captured in ocean gyres: one in the North Atlantic Gyre and the other in the South Pacific Gyre. The plastic "soup mixture" poses health risks to marine animals, fish, and seabirds in all three of these gyres.[58]

Beyond the reach of the gyre patches, plastic is found in all depths and distances of the great seas of Planet Earth. The ocean floor is becoming a trash dump, collecting plastic pollution over the years. A recent study from Australia's national science agency, in partnership with the University of Toronto, estimates that up to 11 million metric tons of plastic pollution is sitting on the ocean floor.[59] "Alice Zhu, a Ph.D. Candidate from the University of Toronto who led the study, said the estimate of plastic pollution on the ocean floor could be up to 100 times more than the amount of plastic floating on the ocean's surface based on recent estimates."[60]

PLASTICS IMPACTING MARINE LIFE

Thousands of marine species—from zooplankton and fish to sea turtles, aquatic mammals, and seabirds—have been observed entangled in plastic debris.[61] Aside from plastic pollution like fishing lines and six-pack plastic holders physically entrapping marine animals, sea birds, and coral reefs,[62] the most dangerous harm to marine life is often unseen. Seabirds and fish consume large quantities of plastics, often confusing trash for food, as many plastics are covered in algae or mixed with seaweed.

Researchers studying seabirds on islands near New Zealand and Australia note that they consume "more plastic as a proportion of their body mass than any other marine animal." In addition, researchers note that the "consumption of plastic just leads to chronic, unrelenting hunger."[63] A marine biologist with the National Oceanic and Atmospheric Administration sums up the confusion: "The unfortunate thing about this is that they're eating plastic thinking it's food; imagine you ate lunch and then just felt weak and lethargic and hungry all day. That would be confusing."[64]

Various studies demonstrate that marine plastics affect the digestive tracts of fish, turtles, and birds, diminishing their urge to eat and altering their feeding behavior, thereby reducing growth and reproductive output. In some cases, the scientists found stomachs stuffed with plastics, leading to starvation.[65]

As the Center for Biological Diversity noted, seabirds often fill their stomachs with wind-blown shredded plastic bags that are mistaken for food. Sea turtles mistake floating plastic shopping bags glaring in sunlight, as they look like jellyfish. Based on a study of over 370 autopsies, one-third of leatherback sea turtles had plastic in their stomachs, most often a plastic bag. In many cases, the plastic blocks the digestive tract and makes the turtle buoyant, making it unable to dive for food.[66]

Within the marine animal's body, the plastic chemicals bioaccumulate and can cause all sorts of disruption. The chemicals can cause excess estrogen to be produced, which has led to the discovery of male fish with female sex organs. A University of Toronto study found that fish that ingested polyethylene particles from San Diego Bay "suffered more liver damage than those that had consumed virgin plastic," indicating that the plastic consumed became a carrier for pollutants that affected the liver.[67] In another experiment, oysters produced fewer eggs and less sperm when ingesting tiny pieces of polystyrene.[68] Ocean fish eat plastic fragments, which are then eaten by larger fish and marine mammals and eventually consumed by humans through the natural food chain.[69]

The quantities of marine life affected by plastics are staggering. According to Ocean Crusaders, approximately one million seabirds and one hundred thousand marine creatures die yearly from plastic entanglement. At least *two-thirds of the world's fish* have ingested plastic.[70] The most frequent artificial item sailors see in the oceans is plastic bags. Marine experts have rated the types of ocean litter most harmful to wildlife, and that list includes plastic fishing nets, fishing gear, balloons, plastic bags, plastic bottle caps, and plastic utensils.[71] Many predict that there could be more plastic in the ocean than marine life (by weight) by 2050.[72]

As plastic deteriorates in the marine environment, it fragments into smaller and more numerous particles, compounding the plastic pollution problem. This fragmentation leads to what are generally called microplastics, which are small enough to be ingested by fish through their gills.

MICROPLASTICS AND NANOPLASTICS

Microplastics are broadly defined as synthetic plastic particles sized smaller than 5 mm that are insoluble in water and not degradable. Nanoplastics are even smaller, with diameters as small as 0.001 mm. There is some misuse of the term *biodegradable* when applied to synthetic fossil fuel-derived plastics. Some even add "millions of years" to the estimated time that plastics will biodegrade in landfills or the natural environment. Some would argue that the existence of microplastics is a process toward biodegradation; however, this displays a misunderstanding of the term and the process.

I borrow a simplified set of definitions for the nonscientific community, yet these definitions hold up to the more complex explanations in scientific journals. There are three important definitions in this discussion:

- **"Biodegradability:** Microbial assimilation of the fragmented product as a food source by the soil microorganisms.

- **Compostability:** Complete assimilation [of the organic material] within 180 days in an industrial compost environment [or in backyard composters].

- **Fragmentation**: Organic matter is broken down into microscopic fragments."[73]

These three definitions are part of the process where organic material will first fragment, then degrade and assimilate, and finally be fully composted. However, biodegrading and composting do not work well for *nonorganic* materials such as plastics. Plastics do fragment and yield microplastics and nanoplastics, yet the resulting material only partially assimilates with the natural environment and becomes a food source for soil animals and microorganisms. Microplastics never biodegrade, even after millions of years—they are here forever and part of the human-made "forever chemistry." The material stays fragmented (does not decompose), becoming a carrier of "floating" chemicals that can become attached to animals and microorganisms through physical, electrostatic, or chemical means. This is an important point, noted in the stories documented next, as microplastics become a carrier of chemical contaminants that impact the health of humans and the entire planet.

Microplastics and nanoplastics are found worldwide, in the natural environment, within all animal species and human bodies. Because the originating plastic product or packaging containing resins may have contained toxic chemicals, the resulting microplastics may carry those toxic chemicals into ecosystems, thus serving as transport vectors. Scientists have documented the trail of microplastics within the general world ecosystems and the human food chain. While their presence is certain, what is largely unknown and a topic of thousands of studies is the health effects on the human body. However, many health experts hypothesize that microplastics are linked to cancers, infertility, and endocrine disruption.

Researchers studying the effects of water and air pollution in U.S. national parks discovered microplastics in their collected rainwater and air samples. They calculated from their samplings that "over 1,000 metric tons of microplastic particles fall from the sky through air pollution into 11 protected areas in the western U.S. each year. That's the equivalent of over 120 million plastic water bottles."[74] The study, published in the *Journal of Science*, notes that microplastics are "blowing all over the world" and "flowing into oceans via wastewater" and now "falling in the form of plastic rain—the new acid rain."[75] Think about that: Microplastics fall as plastic rain and form a new type of acid rain.

Generally, "acid rain" from power plants can be captured by deploying scrubbers, which filter the emissions in the industrial smoke stakes. However, there is no easy capture mechanism for microplastics, as the release is through the breakdown of plastic containers, packaging, and products discarded in the environment. One means to prevent microplastics from entering the environment is to prevent the discarding of plastics into the environment. That is a more significant problem than one can envision. The genie is out of the bottle—plastic waste is expected to skyrocket to 460 million tons annually by 2030.[76]

Snow samples from twenty different regions of Siberia were found to contain plastic microfibers even in remote wilderness areas. Yulia Frank, the science director at Tomsk State University who led this study, stated, "It's clear that it's not just rivers and seas that are involved in circulating microplastics around the world, but also soil, living creatures, and even the atmosphere."[77]

Researchers from the University of Basel have found microplastics on their journey to the Weddell Sea in Antarctica. The study involved "34 surface water samples and 79 subsurface water samples from the isolated Weddell region of Antarctica, filtering over eight million liters of seawater in the process." The researchers wrote, "Being difficult to access due to its year-round sea ice coverage, the Weddell Sea is one of the most remote regions in Antarctica, with particularly low levels of human activity. It plays an important ecological role for endemic species, migrating seabirds, and marine mammals."[78]

Significant microplastic levels were found in samples taken from the snowpack's core at a base camp on Mount Everest, at 27,700 feet elevation. This is the highest elevation in the world where microplastics have been found. The National Geographic Society organized these samples and ten subsequent studies to examine how human actions affect Mount Everest. The analysis indicated that microplastics found atop the mountain comprised polyester, acrylic, nylon, and polypropylene.[79]

Closer to home, consider microplastics in baby bottles. Researchers selected ten types of plastic baby bottles, representing 70 percent of the current marketplace, and prepared infant formula in each based on the World Health Organization guidelines. When heated to 158 degrees, per the instructions, "the bottles released anywhere from 1 million to 16 million particles per liter. The bottles also released trillions of even smaller nanoplastics—tiny bits of plastic ranging in size from 10 nanometers up to 1 micron—so many that we stopped counting them."[80] As the researchers continued their testing of the plastic bottles, microplastics were released for more than twenty-one days from the same bottles without additional heating. The study suggests that increasing the temperature is essential in releasing microplastics, but not the only factor.

As researchers from the State University of New York demonstrated, bottled water has been found to contain microplastics. Through their study, eleven top-selling bottled water brands yielded test results with an average of 325 microplastic pieces per liter. The range varied significantly by brand, from Nestlé Pure Life at 10,390 microplastic particles per liter to San Pellegrino at 74 microplastic particles per liter. The types of plastic identified included rayon and polyethylene. The study's findings suggest that a person who drinks a liter of bottled water daily might consume tens of thousands of microplastic particles yearly.[81]

Extreme caution should be applied when reusing plastics. Sherri Mason, plastic pollution researcher and director of Project NePTWNE at Gannon University, cautions that "single-use plastic water bottle sheds micro- and nanoplastics into your water when you refill it, and a takeout container or frozen meal tray sheds these particles into your food." She also highlights that there are more than "16,000 chemicals found in plastic, over 4,200 of which are considered highly hazardous." The researchers recommend reuse of nonplastic water bottles and use of nonplastic microwave dishes.[82]

Microfibers in our clothing, often woven into sleepwear and other clothing to provide fire resistance or water resistance properties, can also wash down the drain through the washing machine. Washing clothes releases half a million tons of plastic microfibers into the ocean every year, equivalent to more than fifty billion plastic bottles, according to the Ellen MacArthur Foundation.[83] In addition, fabrics shed trace amounts of fibers as they're worn. Synthetic fabrics are measured to shed four hundred microplastic fibers every twenty minutes of use for each *gram* of material, according to researchers in a study.[84]

Microplastics from plastic dishware we eat and drink from can wash down the drain through the dishwasher. In addition, microplastics from our car tires and auto accident debris can also wash down the nearest stormwater drain. As these various water sources drain to our water treatment facilities, microplastics escape the antiquated municipal filtration systems and leach into our natural waterways, perhaps into our drinking water and food systems.

Microbeads, a sphere-shaped microplastic intentionally manufactured for personal care products, can be found in cleansing agents like soaps and washes, skin exfoliants, body creams and moisturizers, hair gel, and toothpaste whiteners. Polyethylene, nylon, polypropylene, and polystyrene are common plastics in microbeads. As with microplastics and microfibers, microbeads wash down the household drain and pass easily through the filters of municipal wastewater facilities into natural bodies of water.

Glitter is made from a combination of aluminum and plastic and is generally classified as microplastic because it often measures less than 5 mm in size per particle. Glitter can be found on greeting cards, birthday balloons, and almost any product related to celebrations. Glitter is also used in some cosmetics and clothing, in school art projects, and craft pieces of all sorts. Unfortunately, as with other microplastics, glitter gets washed down the drain, finds its path into our oceans, and is consumed by aquatic life. *National Geographic* notes that the glitter "bits collect in birds' stomachs, where they can cause them to die of starvation." Researchers and marine scientists continue to study the effects on marine life.[85]

A meta-analysis of fifty studies of seafood yielded startling results about microplastics. The collective evidence involved the study of "four phyla: mollusks, crustaceans, fish, and echinodermata."

Concentrations of microplastics were measured and assessed based on "human uptake from its consumption." The study results noted that "maximum annual human microplastics uptake was estimated to be close to 55,000 MP [microplastics] particles. Statistical, sample and methodological heterogeneity were high." The authors' suggested action was to reduce consumption of seafood to reduce human exposure to microplastics.[86]

But reliance on other protein sources may not solve the problem. Researchers at Ocean Conservancy and the University of Toronto estimate that American adults could be consuming up to 3.8 million microplastics per year from various protein sources. Nearly two thousand adults were studied in this research, and food type, quantity, and serving size were surveyed. Most of the food the researchers acquired from grocers for the study was organic-based and included chicken nuggets, top sirloin steaks, pork loin chops, chicken breasts, plant-based nuggets, plant-based fish sticks, plant-based ground beef, tofu blocks, and breaded shrimp. The study results state that "microplastics were present in all 16 protein products and in 88% of all samples tested (98/111 samples). Six different morphologies of microplastics were observed: fibers, fiber bundles, fragments, foams, and films."[87]

A study by the University of Newcastle, Australia, suggests people consume about two thousand tiny pieces of plastic weekly. Drinking water was the largest source of microplastics, and shellfish was also found to be a large carrier. Marco Lambertini, WWF international director general, said: "These findings must serve as a wake-up call to governments."[88]

Human exposure to microplastics may happen through ingesting contaminated food or drink, inhaling contaminated air, or dermal absorption from contaminated surfaces, clothing, or abrasion. Microplastics "are believed to migrate across body membranes to the gastrointestinal tract, circulatory system, and lungs."[89]

A recent study utilizing a breathing mannequin sitting at a desk in a typical apartment was tested for the inhalation of microplastics from the circulating airflow. The study results found that an average of 11.3 microplastics per hour were inhaled as the mannequin was breathing indoor air in a typical human position. Although this test was done using a mannequin, studying human lungs would be far more complex and challenging. Nevertheless, according to one of the study's authors, "this is the first evidence of human exposure to microplastic through breathing indoor air."[90]

A scientific literature review of the intersection of microplastics and human health offers this observation: "If inhaled or ingested, microplastics may accumulate and exert localized particle toxicity by inducing or enhancing an immune response. Chemical toxicity could occur due to the localized leaching of component monomers, endogenous additives, and adsorbed environmental pollutants. Chronic exposure is anticipated to be of greater concern due to the accumulative effect that could occur."[91] The review recommends much deeper research on exposure levels and mechanisms of toxicity.

In a study of 104 human patients, "nine kinds of microplastics were reported in human body fluids with their size ranging from 19.66 to 103.27 μm." The researchers also noted that microplastic "exposure was unexpectedly high inside the human body despite the protection of biological barriers and membranes, raising awareness of the impact of particle pollution on sustainable development." The study recommendation "urges environmental protection agencies to take the exposure of microparticles including microplastics as risk factors and to evaluate our environment from a novel perspective. Moreover, this particle landscape analysis is anticipated to be a starting point for future funding programs on global particle pollution."[92]

An analysis published by the Program on Reproductive Health & the Environment at the University of California, San Francisco, "suggests fertility problems, neurological diseases, harms to metabolism and the immune system, and changes that signal increased risk of cancer, among other effects" linked to the ingestion of microplastics. Tracey Woodruff, a senior author on the study, notes, "In the field of environmental health, when we have concerning signals, we should be concerned." The authors advise that further research is needed.[93]

As noted previously, microplastics are carriers of toxins. The Global Microplastics Initiative states that pesticides and manufacturing chemicals adhere to microplastics, are carried into oceans, and bioaccumulate in aquatic life. Studies have also shown that these contaminated microplastics move from organisms' digestive tracts into nature's chain. See the Global Microplastics Initiative website for their microplastic tracking database, which offers "the largest and most diverse global microplastic pollution datasets to date."[94]

It should be cause for alarm that humans, including infants, are ingesting, inhaling, and absorbing thousands of toxic microplastics weekly. However, as microplastics are too tiny to see with human eyes, we may not know or feel the magnitude of the plastic problem. This "out of sight, out of mind" mentality keeps people from truly understanding the consequences of plastic pollution. Yet the presence of these pollutants is inarguable and long-lasting, not just for us but also for Earth's climate.

Artwork by Pam Longobardi + Drifters Project

ANXIETY OF APPETITES, 2020

Recovered and assembled ocean-made drift net balls, floats, feathers, barnacles, and bryozoans, 122" x 60" x 60". The factory fishing industry supplies venues like Red Lobster with enormous quantities of ocean life for all-you-can-eat menu items such as "Ultimate Endless Shrimp Now Available All Day, Every Day for a Limited Time starting October 18, 2020." The commercial fishing industry creates a vast amount of marine plastic pollution, primarily from the nets, as well as the myriad wasted lives of bycatch. These ghost nets become tangled and are cut by fishing boats and discarded into the ocean yet continue to hunt, roaming the sea and snaring fish, sharks, seals, seabirds, and whales that die in the nets. The ocean, in an attempt to dispel them, tangles many different nets together and vomits them onto the shore. As fish stocks deplete, there is a growing sense of anxiety among cultures that rely on fishing as their primary means of sustenance.

—PAM LONGOBARDI

Chapter 5

THE INTERSECTION OF PLASTICS AND CLIMATE CHANGE

"If we do not save the environment and save the Earth,
then whatever we do in civil rights or in a war
against poverty will be of no meaning because then
we will have the equality of extinction."

—JAMES L. FARMER[1]

One day, as I was walking the family farm with my grandfather, he picked up a handful of soil and asked me what I saw. As a typical urban fourteen-year-old, I responded, "Dirt." He thought differently. He saw rich, growing soil. It was late April, and he talked to me about the soil's temperature and moisture content. He pulled out a few worms and noted the hidden nitrogen content. Grandfather noted that there was still a risk of a hard freeze and that planting season begins the day after Mother's Day.

I wonder today how many people know the wisdom of the land, as my grandfather did. It may sound like simple mechanics: crop rotation to preserve the soil quality, tilling in a specific pattern to avoid soil erosion and water loss, and choosing the planting and harvesting days carefully around weather patterns. Yet today, so many of us buy food out of season, and grocers ship it halfway around the world to satisfy our tastes. What is our carbon footprint for food purchases today compared to my grandfather's farming practices?

One of my grandfather's last observations to me, as he was aging and retired, was that the spring season was arriving sooner and the summers were hotter, and that was not good news for farmers. Though he did not utilize modern measurement instruments—as a farmer, he was simply an astute observer of nature—his observations were spot-on. Climate scientists have been saying the same: warning of the warming of Earth and the changing growing seasons of our crops due to these warmer seasons.

The Intergovernmental Panel on Climate Change (IPCC) is the United Nations bureau that assesses weather science related to climate change. The IPCC is the world's authority on global warming and climate emergencies, and it collects and evaluates data from scientists worldwide. The stark reality is summed up in this 2018 IPCC statement: "Human activities are estimated to have caused approximately 1.0°C [1.8°F] of global warming above pre-industrial levels, with a likely range of 0.8°C to 1.2°C. Global warming is likely to reach 1.5°C [2.7°F] between 2030 and 2052 if it continues to increase at the current rate."[2]

The importance of this data-driven global warming measurement statement is the warning sign it gives of reaching close to the "tipping point" of 2.0°C (3.6°F) above pre-industrial levels. Some prominent climate scientists, led by Tim Lenton at Exeter University, England, note that this 2.0° C temperature tipping point could activate "a domino-like cascade that could take Earth systems to even higher temperatures."[3] This type of change could eventually lead to an irreversible path toward a "hothouse Earth," as described by Lenton.

A "tipping point" is generally described as a measure at which the activity triggers "cascades" of other damaging activities, eventually causing an *irreversible* shift in the climate to a much hotter world. *Irreversible* is the key word here—if humans are interested in reversing human-induced harm to the climate, now is the time to act before the tipping point is reached.

Climate scientists are discovering many different types of tipping points, such as accelerating sea level rise, burning forests and thawing permafrost, releasing large amounts of carbon dioxide (CO_2), significant changes in the ocean circulation systems, and the global average temperature rise. Moreover, in recent years, scientists have found these measures to be interrelated and changing in concert, warning us that climate changes are happening ever more rapidly than previously predicted.

Other obvious signs of climate change forces in our environment include the following:

- melting of ice in a warmer Arctic, including the melting of permanent glaciers
- melting of permafrost in the Arctic and the Antarctic
- faster melting of mountain snowpack in the spring
- earlier and warmer spring temperatures
- rising sea level due to melting glaciers
- food chain disruption/loss due to warming ocean currents
- warmer global ocean circulation systems
- a shift in heat distribution around the planet
- unusual droughts in certain wet regions (e.g., Amazon rainforest)
- unusual rainfall in certain arid regions
- acceleration in disease development and spread

- extreme fires with tornado-like fire windstorms
- extreme pest infestation
- migratory pattern disruptions
- acceleration of species extinction
- rapid decline in biodiversity
- coral reef die-offs
- changing growing seasons for traditional food crops
- 90 percent global loss of bees, affecting pollination

The list of climate effects grows as the research displays the devastation of Earth from excessive greenhouse gas emissions in the atmosphere, including carbon dioxide (CO2), methane (CH4), nitrous oxide (NO2), and fluorinated gases (F-gases). This book is not written to argue about the existence of climate change—the science is clear, and we are in a climate emergency. In the 25th UN Climate Change Conference (COP25) held in 2019 in Madrid, climate change was described as an "existential threat to civilization." The COP25 conference reports stated that 40.6 billion tons of CO_2 were released into Earth's atmosphere in 2019, an all-time record.[4]

Climate scientists have charted the rise in global CO_2 emitted into the atmosphere and correlated it to the rising global temperature of Earth. The international annual CO_2 ppm (parts per million) concentration increases have been reported through the U.S. Global Change Research Program.[5] For the past seventy years, CO_2 levels have risen 109 parts per million, from 316 in 1959 to 425 in 2024, a 34 percent increase.[6] Over the same span of time, the average global temperature has risen 1.2°C.[7] Climate scientists note significant changes to Earth's physical health and condition when the average global temperature rises more than 1.5°C above post-industrial 1880 temperatures. The

National Oceanic and Atmospheric Association (NOAA) reported in 2024, that "Earth has warmed roughly 1.5°C (2.7°F) above temperatures in the 1800s, before people began burning vast reserves of fossil fuels."[8] The EU's Copernicus Climate Change Service independently measures the warming trends and observed that "February 2025 was the 19th month in a 20-month period for which the global-average surface air temperature was more than 1.5°C above pre-industrial levels."[9]

The corresponding rise in global temperatures alongside the increasing load of atmospheric carbon dioxide (CO_2) is no coincidence. Climate scientists have demonstrated the causal relationship. More to the point of this book, fossil fuel extraction, refinement, and production of plastics have been a significant part of CO_2 emissions, causing that rise in global temperatures.

Reflecting on this global warming, an additional warning note from the IPCC prepares us for the "transformative systemic change" that is necessary to meet this climate emergency:

> Limiting warming to 1.5°C [2.7°F] above pre-industrial levels would require transformative systemic change, integrated with sustainable development. Such change would require the upscaling and acceleration of implementing far-reaching, multilevel, and cross-sectoral climate mitigation and addressing barriers.[10]

The effects of climate change, the need to slow Earth's atmospheric warming, and the need to mitigate the impact of climate change already existing today fall on every nation. However, climate change affects people in different ways and is much more harmful to the world's poor and marginalized.

With sea levels rising, communities living near the water's edge are directly affected. This could be ocean shores as well as islands and peninsulas. It takes monetary resources to relocate your living space—resources the poor cannot access. Some climate scientists estimate that a 51 cm sea rise by 2100 could cause a displacement of upward of 200 million people.[11]

Extreme weather conditions caused by climate change affect food growth, both home-grown and large-scale agriculture. Excessive rain or drought can destroy the growing season, thus causing starvation, extreme malnutrition, and food insecurity in some areas of the world. Infectious diseases can also spread due to disruptions in animal migration and accelerated extinction patterns.

Hurricanes and cyclones can strike areas already devastated by drought to cause extreme flooding, migration, and homelessness. In 2020, the United States suffered sixteen weather events with disaster damage of more than $1 billion *each*.[12] In 2019, Cyclone Idai slammed through a large region in Mozambique, destroying millions of acres of farmland and leaving 1.7 million people needing food assistance.[13]

As of the writing of this book, 2024 has been declared the "hottest year since global temperature records began in 1850."[14] Extreme heat affects those without the aid of air conditioners and cool places of comfort to rest. Marginalized communities often have medical facilities and schools with no air conditioning. In addition, excessive heat can exacerbate many health conditions.

The recent decade's intense heat has dried the forests' timber. Approximately two million acres of California forests burned in 2018, a forty-five-year record according to the U.S. Forest Service. Yet, in 2020, more than four million acres of California trees were burned, and 8,200 homes were destroyed, leaving 53,000 homeless.[15] Australia's Black Summer fire of 2020 destroyed 5.5 million hectares (13.6 million acres) of forest and more than three thousand homes.

As the world nations meet annually to uphold their Paris Agreement pledges, there is a uniform agreement that transformational systemic change is necessary to move away from fossil fuel usage toward alternative energy systems. At the time the Paris Agreement was signed in 2015, Christiana Figueres, the UN's climate chief, stated, "The science is clear: we cannot move backwards to more fossil fuels and more extraction."[16] The goal is to "decarbonize" the world's use of fossil fuels and leave the carbon below ground—no excavation of coal, oil, gas, tar sands, or any fossil fuel sources. Of course, this also implies no plastic production!

But where is the action behind the words? The apparent destruction of Earth and all its varied species by climate change is before us. The nations' pledges are on the table. Still, the physical activities to roll back fossil fuel usage are stalled in political discussions. Unfortunately, the urgency has not been in the wind.

Over the decades, voices have been calling out the need to address climate change, but industry leaders have paid little attention. However, these cries have proven prophetic, as the signs of climate change grow ever more present today for us to witness—as if Earth herself were speaking to us. This is not a political discussion but rather a decision to protect Earth, our natural habitat. It's time to listen to these voices and create action.

THE EARLY VOICES FOR ACTION

The early voices that called out the attention that we have a climate emergency have largely been ignored; however, we can still learn from those voices and seek the wisdom of their visions.

Barry Commoner was an ecologist, a college professor, and one of the founders of the modern environmental movement in the 1960s and '70s. In an interview about climate change, he stated, "If you ask what you are going to do about global warming, the only rational answer is to change how we do transportation, energy production, agriculture, and a good deal of manufacturing. The problem originates in human activity in the production of goods."[17] The manufacturing and production of plastics directly impact our climate, as noted later in this chapter.

In her 1963 book *Silent Spring*, Rachel Carson did not speak of climate change or global warming but of how humans can harm Earth in unimaginable ways. She wrote about the harm caused by the chemical DDT, its indiscriminate spraying on crops and neighborhoods, and how the pesticides containing DDT can enter our food chains and harm human and animal lives. Over time, DDT and other related pesticides were banned worldwide, yet the residual effects remain. Rachel Carson was one of our early environmental pioneers, also credited as one of the founders of the modern environmental movement. The lessons from her teachings can certainly be applied to our concerns about plastic resins in the food chain and the perpetual effects on the environment today.

Although my soul was awakened by reading *Silent Spring*, I became fully aware of the dangers of climate change and the actions caused by humans through the 1989 book *The End of Nature* by Bill McKibben.[18] He wrote about how nature is now being directly and irreversibly affected by the actions of humans. McKibben has written additional books, including *Fight Global Warming Now* (published in 2007),[19] that empower his readers to take action to reverse the effects of climate change. We all can act now by stopping the extraction of fossil fuels and working to decarbonize our economy. This includes the elimination of the production of plastics.

These voices and others were primarily ignored until former vice president Al Gore displayed an alarming global warming trend in the 2006 film *An Inconvenient Truth*.[20] The phrase *planetary emergency* was coined to best describe the need for immediate actions to address the declining climate conditions caused by increasing CO_2 emissions over the last century. Al Gore was granted a Nobel Peace Prize in 2007 for his efforts "to build up and disseminate greater knowledge about man-made climate change and to lay the foundations for the measures needed to counteract such change."[21]

Science journals can leave a cold, hard impression, yet Elizabeth Kolbert offers a different approach in her 2006 book, *Field Notes from a Catastrophe.*[22] Through storytelling, she brings to life the real dangers of climate change as Earth is threatened by the unusual thawing of the Arctic, rising waters, and changing habitats due to extreme weather patterns. The vignettes are compelling, even while exposing the politics of climate change. Kolbert notes that we are approaching "a critical threshold in terms of Earth's climate"—a threshold that may be impossible to turn back from.

Jacques Cousteau was the oceanic hero of my childhood and the most famous underwater explorer of recent history. As his ocean explorations found significant amounts of human-generated trash, including plastics, he spent the last twenty-five years of his life dedicated to raising awareness about human intervention in underwater ecosystems. Cousteau is commonly attributed as having said, "Water and air, the two essential fluids on which all life depends, have become global garbage cans." To further his awareness campaign, he founded the Cousteau Society in 1973 and made documentary feature films with an environmental theme to protect ocean habitats.[23]

Many other voices from the past tell the same story of our throwaway culture and its devastating effects on Planet Earth. Yet this book doesn't have the space to include the wisdom offered from the past centuries, as even the indigenous Americans who lived here first watched the European settlers trash and desecrate the beautiful land, rivers, and seas. So, the question at hand is: Why did we not listen to these wise voices?

COGNITIVE DISSONANCE

Several centuries ago, most people thought Earth was flat until science proved that Earth was a sphere. However, did you know that Greek philosophers wrote the first concepts of a spherical Earth in the fifth century BC?[24] Through the following centuries, many various ancient civilizations attempted to prove the theory correct, up to a physical demonstration in 1519 by Ferdinand Magellan and Juan Sebastián del Cano with their voyage around the world.[25] Most of the world's population believed in a "round" Earth by the nineteenth century, perhaps from school teachings. Yet a significant amount of Earth's population—the Flat Earthers—did not believe in a round Earth until 1969 when the Apollo spacecraft mission displayed photos of Earth from space. As you can see, it is difficult to change our past belief systems.

This "disbelief" of a physical fact is often attributed to *cognitive dissonance*: when a person can hold contradictory beliefs, ideas, or values that are not consistent with data or general information made available to that person, if that person internally evaluates a conflict of information and tries to reduce the "stress" by taking sides, usually providing a defense of a past belief or idea rather than embracing a new belief or idea.[26] This conflict resolution process—taking the "familiar" side of an issue—eliminates the individual's cognitive dissonance. Yet that individual may be out of step with the rest of society, which is moving on to a new reality with a new set of beliefs and facts that others are embracing. In our fast-moving world, it can be hard to discern truth from fiction; cognitive dissonance can be a method to settle the argument in our brain, but it's an unconscious decision, not a conscious fact-finding rational means to sort out fact from fiction.

Psychologist Leon Festinger coined the term *cognitive dissonance* and provided much research behind the concept. His research pointed to three reasons why people engage in an unconscious decision-making process that involves cognitive dissonance:

- "The change may be painful or involve loss.
- The present behavior may be otherwise satisfying.

- Making the change may simply not be possible."[27]

According to Festinger, our brain's goal is to seek consistency within our beliefs or past conceptual understandings. Therefore, new beliefs and concepts must overcome this unconscious bias. Humans are subject to this form of nonrational decision-making, yet we must know its dangers and potential pitfalls. We can miss opportunities for change and make poor judgments by ignoring the available rational, analytical decision tools.

An example to consider is the small portion of the population who disbelieve in climate change. Those caught up with cognitive dissonance decision-making may not be able to embrace the science of climate change and the reality of Earth's changing ecosystems due to destructive human forces. This book is not charged with convincing those climate deniers to change their minds. However, it is essential to understand how our friends, family, and neighbors react to the topic of climate change, and cognitive dissonance may be part of the scene. It is a potential explanation for why some people argue that climate change does not exist. Or if they do recognize climate change, their position might be that it is not induced by human activity. Over the past few decades, incontrovertible proof has mounted from hypothesis to correlation to a direct causal relationship between human forces and climate change. Yet that is not a convincing argument for some—displaying the powerful forces of cognitive dissonance.

The preceding example is what is characterized as a *climate denier*. Now, consider an example of an environmentalist who does believe in this climate emergency and works hard to reverse the effects of climate change, and yet is equally subject to the forces of cognitive dissonance without their knowledge. This book calls for readers to change their thinking and lifestyles. That's no small thing! When a new external set of information requires a significant difference or change in how we live, and the decision-making process is filtered through cognitive dissonance, the new lifestyle will likely not be chosen, as that would be the difficult and complex path. In this respect, the climate denier and the environmentalist are both working against the same human tendencies.

But what if we were serious about this climate emergency—serious enough to make significant greenhouse gas (GHG) reductions to reach carbon neutrality? To achieve that carbon neutrality, as necessary to prevent a breach of Earth's temperature tipping point, the Paris Agreements (COP21) call upon the complete conversion of the fossil fuel industry toward renewable energy sources.[28]

An immediate takeaway from this for an environmentalist is to start researching electric/battery-powered modes of transportation, solarizing one's house, and learning how to move their community from a fossil fuel electric grid to a renewable energy grid. Notice that the focal point is on the traditional uses of fossil fuels: transportation and electrical generation. Yet 12 percent of fossil fuel industrial production is geared toward the production of plastics, and that segment is likely to double in the next two decades! Are you ready to give up all plastics—not just single-use plastics but *all* plastics—to reach the goal of a fossil-fuel-free society? If you feel resistance, that may be the cognitive dissonance reasoning in your head trying to wiggle you out of the lifestyle-change commitment. After all, going entirely fossil fuel-free would be a lifestyle change for everyone.

Are you ready for such a new lifestyle? Will your self-justification be "I already recycle, and I compost, and I've done my part" or "I live a green and environmentally sound lifestyle with a reduced carbon footprint; I've gone the distance; now others need to pick up the pace"? Does it sound like cognitive dissonance reasoning rather than a rational means of facing a lifestyle decision? Some will go the distance and say plastics are not related to fossil fuels after they are produced and do not impact climate change. Suppose your "reasoning" implies that conclusion—read on for a more factual demonstration of the connection between plastics and fossil fuels. As you read, please keep in check that immediate human impulse to turn down the easy path of least resistance through cognitive dissonance reasoning.

FOSSIL FUEL EXTRACTION AND TRANSPORT

All fossil fuels require some form of resource extraction from Earth. The main types of fossil fuels are coal, petroleum, natural gas, oil shale, natural bitumen, tar sands, and heavy oils, all of which are rich in hydrocarbons.[29] These fossil fuels were formed through an ancient decay process of organic material such as algae and bacteria that began in the Archean Eon (4 billion years ago) and from plants as early as the Devonian Period (400 million years ago).[30]

The primary forms of extraction from Earth are mining and fracking. The industrial age brought about the human need for intensive amounts of fossil fuels, causing significant increases in land mining extraction methods, with hydraulic fracking being a more recent method. By-products of the extraction method are carbon dioxide (CO_2) and methane (CH_4) gas release emissions, both classified as greenhouse gases.

Notice the innocuous industry term for these unmitigated and uncaptured gas releases: *by-products*. An industrial by-product is generally defined as something produced in addition to the principal product and marketed as an additional product line. However, in this instance, the petroleum industry does not typically capture CO_2 and CH_4 from their points of origin due to how extremely expensive it would be, unless a governmental agency requires them to do so. Therefore, I question the usage of the term *by-products* and prefer the term *pollutants*, as the emissions are released as unmitigated and uncaptured gas into the atmosphere.

These gases—pollutants—are released at the extraction sites, in the land clearing, and in the actual extraction work on strip mining and deep coal mines. Emissions are released from the gas flaring devices and exposed gas and oil vaporization at flaring sites. Hundreds of chemicals are utilized at hydraulic fracking sites, each with a vaporization rate that exposes workers and the atmosphere to chemicals. The well pads, the pipelines, and the tanker trucks have leaks in their loading systems, all with atmospheric exposure to greenhouse emissions.

According to the Plastic & Climate report by the Center for International Environmental Law (CIEL), emissions from fossil fuel extraction and production attributed directly to the production of plastics were calculated to be at least 9.5 to 10.5 million metric tons of CO_2 equivalents (CO_2e) in 2015.[31] For extraction and refining dedicated to plastic production on a global scale, approximately 128 million metric tons of CO_2e were calculated for 2015.[32]

PLASTIC REFINING AND MANUFACTURE

Oil and gas refining is the third largest greenhouse gas (GHG)-emitting industrial sector among stationary sources. In 2013, 145 facilities within the United States in the petroleum refineries sector reported GHG emissions of 176.7 million metric tons of CO_2e.[33] Some of this oil and gas is then dedicated to manufacturing and producing plastics. In addition, there are dedicated plastic refining production sites within the fossil fuel industry facilities—a form of vertical integration of the full production and manufacturing of plastics.

The dedicated plastics facilities are very energy intensive, consuming oil or gas to operate the facilities themselves, which means operating emissions. In addition, after hydraulic fracking, the petrol products require significant energy to refine through a process known as cracking of the alkanes by polymerizing olefins into plastic resins.

In 2015, twenty-four ethylene facilities in the United States produced 17.5 million metric tons of CO_2e, while globally, the 2015 cracking emissions to produce plastics was approximately 200 million metric tons of CO_2e.[34]

An additional 349 new petrochemical projects have been constructed or planned in the United States since 2010, most dedicated to plastic production—even in the wake of declining oil and gas profits and the ongoing climate emergency discussions.[35] These facilities create air and water pollution in the surrounding communities and require the fracked gas infrastructure to supply the fossil fuel feedstock. Moreover, these new facilities add new CO_2 emissions that contribute to our climate emergency and the warming of Earth's atmosphere. Based on current growth trends, annual CO_2 emissions from plastic production and incineration are expected to grow to nearly three billion metric tons.[36]

PLASTICS' ONGOING IMPACT

As we have discussed in this book—and witnessed in our lives—plastic is utilized in all facets of our modern world. However, plastic products and packaging, once manufactured and in use for their intended purpose, also produce varying amounts of impacts on our environment and Planet Earth. As discussed in Chapter 4, the impacts on humans, animals, the waterways and oceans, and throughout the natural world are pervasive.

Defenders of the plastics industry will claim that the issue is the misuse of plastics, yet I will note that plastic products and packaging do not come with operating instructions on best-use practices. Furthermore, the issue is more about overproduction, excessive use, and reliance on plastics when less impactful alternatives are available. The actual consumer use of plastics drives up demand for the production of more plastics, justifying the capital funding for more ethylene cracker facilities, which in turn drives up the demand for more hydraulic fracking facilities and pipelines. I don't see anyone from the plastics industry on the frontline trying to discourage sales or offer best-use instructions.

Suppose plastic products are intended to be durable and have an extended life. In that case, the worn-out plastic items would have no end-of-life destination such as reuse, rehab, or disassembly systems as one might find in an actual circular economy, as the item is meant to last. This is not the case for most plastics, which are marketed as disposable or semi-disposable. In the United States, plastic disposal happens through landfills (75.4 percent), incineration (15.3 percent), and recycling (9.3 percent), not accounting for plastic litter.[37] Each disposal method emits greenhouse gases, mainly carbon dioxide (CO_2) and, in some cases, methane (CH_4). Recycling collection and processing does release CO_2 emissions, but offers to offset positive markers by displacing new virgin plastic materials and using less electricity than new material feedstock—yet only about 9 percent of overall plastics in circulation are actually recycled.

On the other side of the scale, incineration has the highest *measurable* emissions regarding plastics' end-of-life. Again, this does not account for the unmeasurable environmental impacts of plastics littered worldwide, primarily plastic packaging, and the significant volume of unrecyclable plastic products. Many countries utilize incineration as part of their "recycling" platform, although incineration is a disposal method and not a recycling method. In 2015, the U.S. emissions from plastic incineration were approximately 5.9 million metric tons of CO_2e, and globally, approximately 16 million metric tons of CO_2e.[38]

Much of the remaining plastics are disposed of "unmanaged" through litter, illegal dumping, open burning, or illegal storage from poor recycling structures. In short, plastics are mismanaged and create environmental damage throughout the globe, including releasing CO_2 emissions at every stage of production, use, and disposal.

A new study by University College London "estimates that air pollution from burning fossil fuels caused 8.7 million premature deaths worldwide in 2018."[39] This fossil fuel impact is from burning emissions credited to "power plants and other emissions sources, including vehicles, trains, diesel generators, and coal used in homes."[40] Yet, also included in the life cycle generation of these quoted emissions is fossil fuel production to produce plastics, the direct production of plastics, the manufacture of plastics products and packaging, the transportation and retailing of plastics, and the consumption and use of plastics— all of which cause premature deaths, according to this medical study, as well as additional harm to human health. Do we label this medical harm *collateral damage* as we love our plastics addiction, or are we willing to recognize the direct linkage of plastics to human harm?

Studies are underway to determine the harm plastics cause to our water quality, which can be difficult to measure. Some would argue that dilution is the solution, but those involved in water quality will vehemently argue against that approach. Nevertheless, plastics are so pervasive in streams, lakes, and oceans that no dilution solution exists to discuss. As noted earlier, microplastics cover the entire planet, including all the water systems, no matter how remote. It now appears that microplastics may interfere with the ocean's capacity to absorb and sequester carbon dioxide. As oceans have historically been a reliable "carbon sink," this discovery is highly problematic and may accelerate CO_2 atmospheric releases.[41]

Closer to home, wastewater treatment facilities and freshwater filtration facilities rely on microfilters to remove contaminates from the water systems. However, as the water inlets have ever-increasing flows of microplastics, the filtration systems cause the microplastics to collide with one another, which then causes them to fragment into nanoplastics. These nanoplastics block pores in the filtration systems, causing damage to the water treatment plants and wearing down the treatment units. This reduces the plants' water flow and operational efficiency and increases the risks that treated water may not meet the required safety standards.[42]

A study released in 2024 found that "the processes of the *entire Earth system*" were changing as a result of plastics pollution, affecting "all pressing global environmental problems, including climate change, biodiversity loss, ocean acidification, and the use of freshwater and land." The lead author, Patricia Villarrubia-Gómez of Stockholm Resilience Centre, stated, "It's necessary to consider the full life cycle of plastics, starting from the extraction of fossil fuel and the primary plastic polymer production."[43]

PLASTICS' PROJECTED GROWTH CURVE

As the attack on fossil fuel transportation methods and forms of electricity supply is going at full throttle, there is intense pressure to convert to greener energy sources. The legislative and consumer focus is on converting vehicle fleets and residential electrical choices—what we might label a *decarbonization* effort. What is missing from these conversations is the plastics industry's future plans to expand its marketing of plastics. Yes, even as drivers and consumers "go green," the fossil fuel industry is protecting its future by an entire market swing toward embracing and enlarging its plastics development schemes! These expansion plans are not held in secret, as the land purchases for the new cracker plants are brought up in public hearings with the proclamation of new jobs for poor communities. As of February 2021, the American Chemistry Council reports that 349 chemical industry investment projects are underway, including new facilities, expansions, and factory restarts.[44]

In recent years, there has been a sharp integration of plastics production companies within the fossil fuel supply chain—an investment into future markets as the oil and gas markets are declining. As a single example, ExxonMobil, the world's largest investor-owned fossil fuel company, owns Exxon Chemical, which produces plastic resins and products, both directly and through subsidiaries. Other examples include Shell, Chevron Phillips, and Total, all of which have invested significantly in plastics infrastructure.

This integration investment is a hedge for fossil fuel companies and a glimpse into the future of their marketing plans: sell more plastics to consumers. As sales may decline in gas and oil markets in the coming years, they are banking on increasing sales and profits in their chemical marketplaces—primarily plastics. For example, Exxon Chemical brought in approximately 10 percent of ExxonMobil's overall revenues and more than 25 percent of its 2015 annual profits.[45] And its investor prospectus indicates these profits will increase as plastics industry investments increase in the coming years. Some reports note that with the projected number of fracking facilities and cracker plants under construction or proposed in 2018, plastics production will double by 2030.[46]

Putting these growth projections into global emissions, the Center for International Environmental Law (CIEL) estimates, "If the production, disposal, and incineration of plastic continue on their present growth trajectory, by 2030, these global emissions could reach 1.34 gigatons per year."[47] Their Plastic & Climate report notes, with a clear demonstration of facts, that the "plastic and petrochemical industries' plans to expand plastic production threaten to exacerbate plastic's climate impacts and *could make limiting global temperature rise to 1.5°C impossible.*"[48]

As greenhouse gases are released into the atmosphere, there is a cumulative effect over time, not simply on an arithmetic or geometric scale but rather by a more complex algorithm with many moving parts in nature. For example, exacerbating elements, such as methane (CH_4) trapped beneath the polar ice caps, are released to the atmosphere in pockets as the ice melts. The released methane then accelerates the warming of the atmosphere beyond current calculations. The point is that the growth projections of the plastics industries and their direct impact on global emissions may very well bring us to one of those tipping points of no return. Intentionally or unintentionally—that is beside the point. Intentions are not up for debate—*actions toward increasing or reducing our impact on the warming of the atmosphere are all that really matter.*

As the authors of The Plastics & Climate Project so rightly put it, *"We are at risk of heating the planet to uninhabitable levels by producing superfluous, disposable packaging that we simply do not need."*[49]

Artwork by Pam Longobardi + Drifters Project

INSTALLATION VIEW OF *ANCHOR/ALBATROSS* IN *OCEAN GLEANING*, 2023

This is a part of the extensive solo exhibition *Ocean Gleaning* exhibited at the Baker Museum Artis–Naples in Naples, Florida, presenting a fifteen-year survey of my Drifters Project work.

—PAM LONGOBARDI

Chapter 6

CHALLENGES TO RECYCLING PLASTICS

"As we look ahead, we need to strive for an environment, and a civilization, able to handle unexpected changes without threatening to collapse. Such a world would be more than simply sustainable; it would be regenerative and diverse, relying on the capacity not only to absorb shocks like the popped housing bubble or rising sea levels, but to evolve with them. In a word, it would be resilient."

—JAMAIS CASCIO[1]

I have spent the vast majority of my career promoting recycling, including the recycling of plastics. I started my recycling career as a college student at the University of Cincinnati, redeveloping the Cincinnati Experience campus recycling drop-off during the late 1970s based on the successful community-supported recycling drop-off models exemplified by Ann Arbor's Ecology Center/Recycle Ann Arbor (led by Mike Garfield, Dan Ezekiel, and others) and the Champaign-Urbana Community Recycling Center (organized by Steve Apotheker). We expanded the Cincinnati program to eleven community drop-offs, serviced weekly by community organizations. Unfortunately, the program folded in 1983, but the life and spirit of the recyclers from the community taught me valuable key lessons that carried forward into an enriching recycling career.

As my recycling career continued, I experimented with various collection methods, from community drop-offs to cash buy-back centers to multi-bin (source-separated) curbside pickup—where the residents separated items into different color-coded bins. When single-stream recycling collection first arrived—where residents placed all recyclables into one container to be sorted at the recycling facility—I protested because it added significant cost and cross-contamination to the recycling program. The primary benefit to single-stream recycling was the added number of converts that would try to support recycling in their homes—yet I continued to be a skeptic for the added trash flows and contamination it added to the recycling streams. What once was a 98 percent recycling stream quickly became 75 percent recycling and 25 percent trash; the resident or local government had to pay someone to sort out the trash and pay to dispose of it—volunteers do not work with waste. I eventually "managed" two single-stream recycling programs over the course of my career: the City of Fresno and the City of Austin. Both experiences strongly confirmed my distaste for single-stream recycling—again for its cross-contamination and added costs.

Quickly, I learned that public education was vital to recycling and the various modes of educational awareness and tools necessary to inform recyclers of the proper way to recycle and to motivate others to join the effort. It has been a lifetime of learning what motivates the average citizen to recycle and how to get residents to read the instructions.

One of the costliest items to sort in a single-stream recycling program is plastics. So many different types and shapes of plastics are coming through the system, requiring manual labor, automated pickers, and robotics to sort the plastics into purer streams. Depending on the local markets, a significant amount of the incoming plastics is sent to the dumpster to be landfilled, as it does not fit the marketing parameters of the local recycling program. The most significant educational challenge of any recycling program—especially a single-stream recycling collection program—is the residents' education on the type of "acceptable plastics" the program can recycle. There is always confusion about what plastics are recyclable and which are unacceptable in any recycling program. "Wish-recycling" often happens, where residents add nonrecyclable plastics to their recycling bin, wishing the recycling program would recycle the plastic items. Unfortunately, this wish-recycling only adds cost to sorting out the added nonrecyclable plastics and placing them in the trash load.

CULTURAL IDENTITY: THE UNIVERSAL PLASTICS AGE

Beyond the cost it presents to recycling programs, there is another aspect to consider with the wish-recycling of plastics: the cultural identity that Americans have placed on plastics within the Universal Plastics Age— the current cohort of generations is consuming plastics at an ever-faster pace, as it is now the normative base material for this society.

Metal and wood were the normative base materials for the Industrial Age. At the same time, silica sand (quartz) and heavy metals (e.g., copper, gold, aluminum, zinc, iron, and nickel) are the material bases for the Computer Age. The era I refer to as the Universal Plastics Age began roughly in the 1970s, after a half-century of plastic experimentation in the furniture and toy markets and the beginning phases of plastics fascination. The 1970s brought plastics into every facet of consumer life and integrated the single-use concept with leisure, ease of use, and a carefree lifestyle. I've labeled this third wave of plastics *Universal* simply because plastics now invade every aspect of our lives, every home, every office, every school, and every part of our daily lives. (See Chapter 2 for further explanation.) Plastics became the normative base material of the current age, without question, thus driving the wish-recycling movement that recyclers fight worldwide.

An illustrative point is the example of the single-use water bottle. The 16-ounce water bottle originated in the 2-liter classic PET soda bottle that began its life in 1973 as a nonrecyclable two-part bottle. As noted in Chapter 2, a financial grant through the assistance of the National Recycling Coalition awarded a research facility lab the task of redesigning the 2-liter bottle into a fully recyclable bottle. The success of that project was translated into the recyclable design of the smaller single-use water bottle. Was it a remarkable breakthrough for recycling? Consider this: *In 2018, Americans bought more than 70 billion plastic water bottles of one gallon or less, with three out of four ending up in a landfill or incinerator.*[2] Why? There is no single good answer to that question. The growth in the use of single-use water bottles is astonishing. According to the CRI, "plastic water bottle sales increased by 2,767% from 1997 (3 billion units) to 2021 (86 billion units)."[3] Another question we should ask is why we are shipping water bottles worldwide when Americans have ready access to tap water. I get the concern about freshwater needs where there is no fresh water. I get the need for bottled water where there are lead pipes, but those settings do not account for the billions of bottles purchased and utilized. Do we not consider waste reduction in our purchasing habits of single-use plastic bottles?

The world of plastics is part of our American cultural identity, an extension of the "disposable society" represented so aptly by the single-use plastics we use once and then throw away. Sally Lee describes the cultural link in her book entitled *The Throwaway Society*, noting the causes of population growth, American wealth, excessive packaging, and a careless attitude of throwing it "away"—somewhere—where we do not need to care about it.[4] The throwaway cultural aspect of plastics is so interwoven into our lifestyles that it is second nature—we toss things in the trash without a guilty thought that anything is wrong with this picture. A great awakening is necessary for people to see that there is a significant amount of harm to this blind, excessive use of plastics.

We must also realize that looking ahead, the remaining driving force of the fossil fuel industry is the production of plastic products and packaging that consumers and businesses utilize, then throw away somewhere. The production of plastics is anticipated to drive "more than a third of the growth in world oil demand by 2030, and half of all growth by 2050, according to the International Energy Agency."[5] This will undoubtedly drive higher carbon emissions, resulting in increasingly warmer global temperatures and a harder edge to climate change. It is all interconnected.

The great plastics awakening must happen soon.

CULTURAL IDENTITY: A REVISIT TO AMERICAN RECYCLING

Another American cultural identity is that of recycling—we identify with the environmental spirit of recycling. Recycling is often said to be the first point-of-entry for a family to explore reducing its environmental impact. Other environmental entry points might be water conservation, energy reduction, and composting. Recycling has a staying power with children fascinated by how we can make new products from old discards. I spent a long career attempting to make recycling a mainstream American activity. I'm proud of the recycling progress I've seen in American homes—and disappointed at the plastic recycling industry's pitfalls. (Sidenote: While there is a significant problem with plastics, it does not mean the rest of our recycling system is broken.)

The cultural identity of recycling is so strong that it offers many benefits, but some cultural pathways (e.g., plastics) are leading us astray. For example, recycling reduces energy and water usage by avoiding the mining of new raw materials, offering lower environmental impacts than landfilling or other disposal methods. There are many reasons to recycle, and I encourage every reader to continue that daily practice at home, at school, in the office, and elsewhere.

PLASTIC RECYCLING "GREENWASHING"

Residents send too many nonrecyclable plastics through their local recycling programs. The misleading plastic codes—noted in Chapter 3—often lead to an overwhelming amount of nonrecyclable plastics being delivered to the regional material recovery facility (MRF). A recent plastics survey reports, "U.S. companies are incorrectly labeling many plastic products as recyclable."[6]

A classic example of a product with a misleading plastic code is the single-use plastic coffee pod, marked with a #5 PP. There are many manufactures of coffee pods; however, the leading brand owner attempted to advertise the recyclability of these pods. Keurig claimed in 2019 and 2020 that its K-Cup pods "can be effectively recycled." However, in court disclosures, "Keurig did not disclose that two of the largest recycling companies in the United States had expressed significant concerns to Keurig regarding the commercial feasibility of curbside recycling of K-Cup pods." The Securities and Exchange Commission (SEC) charged Keurig with false recyclability claims, or greenwashing. The claims were settled with a $1.5 million civil penalty.[7] Despite this large civil fine, many coffee pods on the market still display the recycling chasing arrows, causing residents to mistakenly place them in their recycling bins.

This overburden of unwelcome plastics creates added financial costs to the recycling system. The MRF operator needs to hire human sorters and purchase sorting machinery to glean out the various types of plastics received, whether there is a ready market or not. Film plastics are a particular nuisance because they can clog up the mechanical systems of the recycling MRF, causing downtime when workers knife out the plastic film from the equipment. Those hours of downtime can add unnecessary costs to the recycling program and can be avoided if residents avoid placing film plastics in their recycling bins. In addition to added labor and machinery costs, extra floor space is required in an MRF for all this extra plastics handling, which comes at a price tag as well through the capital investment of the facility and ongoing maintenance. Overall, through a material flow analysis of all traditional residential recyclables, plastics carry a substantial environmental footprint in the recycling collection and processing network.

We can't turn a blind eye to our overconsumption of plastics with the motto that we can recycle them. Because the truth is, we cannot recycle our way out of this mess. The overproduction of plastics and the overconsumption in our lifestyles of plastics is following an unsustainable upward curve straining Earth's carrying capacity, as measured through fossil fuel consumption and the rise of Earth's global temperature. Don't be blinded by plastic recycling's star power: Fossil fuel usage is linked to climate change, and plastics production is directly linked to increased fossil fuel production.

EXPORTING PLASTICS AND THE CHINA CRISIS

The lessons learned from exporting recyclables to China seem to be a well-told story in the news lately, but much is hidden behind the scenes. In addition, the American recycling industry has learned a hard lesson that requires significant rebuilding of the lost recycling infrastructure domestically.

America's modern recycling programs sprouted up organically in major cities, and eventually in rural communities. Throughout the last three decades of the last century, the general format was to collect and sort the recyclables and deliver the recyclables to a market within two hundred miles. This market structure was severely limited, causing a very slow growth of recycling in the United States compared to similar programs in Europe.

To expand recycling markets, in 1995 the Chicago Board of Trade offered—for the first time in history—the electronic commodities trading of recyclables between buyers and sellers.[8] It managed to do so through a collaborative effort with the National Recycling Coalition, the USEPA, the Clean Washington Center, and the New York State Office of Recycling Market Development. This first trading session was the innocent days of "going global," as buyers could electronically buy and trade recyclables just like any other "commodity." With that move, recycling matured and started to be a player in the global marketplace.

As recycling collection programs grew in the early 2000s, specific local markets found it more attractive to ship their products abroad rather than stay domestic. This is partly because of the shipping rate structures within the United States, through over-the-road trucking rates and rail rates compared to the lower ocean liner shipping rates. In addition, low ocean back-haul rates became very attractive, as ocean freighters carried consumer goods to America and empty containers needed to be shipped back to their home ports. The back-haul container rates were incentives to utilize the space in the ships heading back primarily to China and Southeast Asia. Furthermore, innovation in the transportation and logistics markets reduced the number of forklift transfers of loads through the invention of the intermodal container (ICU) or shipping container that can be transferred from truck to rail to ship. These incentives and changes in the shipping markets encouraged recyclers to ship internationally.

In the 1990s and early 2000s, China and much of Southeast Asia were in a robust economic revival, attempting to build for a growing population. From a World Bank report: "Since China began to open up and reform its economy in 1978, GDP growth has averaged almost 10 percent a year, and more than 800 million people have been lifted out of poverty."[9]

China's economic growth opened new markets for American baled plastics. The general purchase plan was #1 PET bales, #2 HDPE bales, and "mixed plastic" #3-7 bales. There were other purchase plans, but this "mixed plastic" plan was very appealing and provided a market outlet unavailable in the United States. These buyers also provided back-hauling shipping deals—far lower shipping rates than truck hauling. Did anyone ask what the plastics' final disposition after processing in China was? Some did but only received vague answers. A few visited China and were impressed with its extensive processing capabilities. Only later did we learn about the environmental detriments to our casual disposal of our mislabeled, miscellaneous plastics, essentially passing off the problem.

As China's economic growth continued, its leaders faced environmental concerns: air, water, and land pollution. China has become the most significant contributor to climate change, as ranked by country.[10] In efforts to get a handle on this problem, China formed its Ministry of Ecology and Environment in 2018 and has aggressively dealt with "imported pollution" (e.g., plastics) to clean up China.[11]

From a recycling perspective, good things are happening within China, but with adverse ripple effects on the American recycling system. China first began tightening quality standards with the Green Fence policy in February 2013, providing the first significant warning to America's recycling industry. Americans had gotten used to "wish-recycling," piling anything we want into the recycling bin and wishing it to be recycled, and it was dirtying up the recycling streams flowing into China.

However, these restrictions were largely ignored. Then, in February 2017, China lowered the hammer with the National Sword import restrictions and enforcement provisions. The National Recycling Coalition[12] and Institute of Scrap Recycling Industries (ISRI) in the United States protested to the World Trade Organization, but the limits were allowed to be implemented. Finally, in March 2018, China implemented BlueSky as an additional set of import restrictions on recyclable commodities entering from American ports . In July 2018, China expressed its intention to implement a total recyclable import ban as of 2020.[13]

China objected in 2013, and America did not clean up its recycling stream. China objected more loudly in 2017 and 2018, and the American recycling industry started to cry foul, but American residents did not change their bad habits. That is partly because the plastics industry still conveys the message that all plastics are recyclable when they are not. As a result, we are polluting our national recycling systems, just as we polluted China's recycling system.

Meanwhile, the construction industry in China was building its own recycling infrastructure, replacing the recyclables once imported from America and the European Union. China no longer needed America's recyclables nor the trash that came with them. With China and much of Southeast Asia saying no more to our recyclables, why can't we turn on the spigot and recycle our plastics here in America? Of course, part of the answer would require us to clean up the trash and wish-recycling from our recycling streams. However, the problem is more significant than that. An excellent 2017 study released by More Recycling documents first that there is "insufficient domestic demand for post-consumer resin (PCR)" and also notes that domestic plastic convertors "could process only 76 percent of the PE PCR acquired for recycling in the U.S."[14]

Let's break down those two components. First, the report notes that Americans are not buying enough products made from recycled content and not demonstrating their consumer preference for recycled content to manufacturers through their purchasing power. In other words, there is not enough consumer demand for recycled content goods; thus, the ripple effect is fewer recycled plastics being purchased by manufacturers. Second, the American recycling infrastructure is limited in its capacity to process PET (polyethylene, e.g., 2-liter bottles), thus putting an upper cap on the amount of PE plastic that can be recycled into products within the United States. The supply and demand sides of the American recycling market were broken and needed repair.

There is much more to say about the recycling industry's challenges toward recycling plastics. Not all industry experts agree on a singular direction. Most are in the mode of trying to fix the system. Yet, in any "fix" when it comes to recycling, the plastics coding system is a significant part of the problem. The SPI plastic coding causes more confusion that anything else and does not assist recycling programs or recycling MRF operators as they try to communicate what "marketable plastic" is. We can begin with eliminating the consumer source of confusion. In its place, there needs to be clear and concise public education about what is and is not recyclable. That is the first and most important step to cleaning up our homegrown recycling steams.

"SINGLE-USE" PLASTICS AND "DURABLE GOODS" PLASTICS

Single-use plastics are under attack by most environmental organizations, and for many good reasons. Single-use plastic bags have an estimated life expectancy of twelve minutes of beneficial primary use, and then they are disposed of.[15] If it's a straw or drink cup, it might be helpful for the time you take to consume the drink, then you toss it in a trash can or the litter stream. If it's a grocery or retail plastic bag, it might carry your groceries home, then become a trash can liner or a pet waste bag—which is not "recycling," as some would say, but rather a pathway to disposal. According to the most generous industry estimates, 10 percent of plastic bags are recycled. Zero percent of the other single-use plastics will find their way through a recycling system.

Single-use plastics, by definition, include product packaging and products utilized only once. Product packaging as a larger category may consist of paper products, foils, glass, metals, and plastics. The global packaging market is estimated to gross $700 billion annually, with a growth rate of 5.6 percent.[16] For this discussion, we will focus just on plastic packaging. When categorizing plastic waste generation by industry type, "product packaging" rises to the top, accounting for 59 percent of all plastic waste by weight in Europe and 65 percent in the United States.[17]

Consider plastic clamshells and blister packs, hinged plastic shells designed to protect retail or food products from harm or theft. You may need a knife or scissors to remove the product from the packaging, which is made from HDPE. It is not likely to be recyclable in your local recycling program. If it can be recycled locally, the likelihood that residents are recycling plastic clamshells or blister packs is extremely low. USEPA recycling statistics reflect zero recycling of plastic clamshells.

Single-serving beverage containers, which are by definition single-use, carry a significant burden of plastic pollution and are the focus of public pressure on beverage companies to capture and recycle as high a percentage as possible. According to the Container Recycling Institute, "15.3 million metric tons of carbon dioxide ($MTCO_2$) are emitted each year in the U.S. to replace wasted beverage containers."[18] Therefore, it was disappointing news throughout the recycling industry when Coca-Cola explained in a press release it now aims "to use 35% to 40% recycled material in primary packaging" by an undetermined set goal date—a drastic reduction from its previous goal of 50% by 2030.[19]

Plastic single-serving bottles are a large but unknown subset of the "primary packaging" mentioned in this press release.

Of additional note are the various litter pickup statistics demonstrating the latest trends in human littering habits. In 2019, an annual one-day coordinated litter pickup in 116 countries captured 32.5 million items. In categorizing items in this litter pickup, plastic food wrappers topped the list, with 4.7 million individual wrappers collected and counted.[20] In my youth, candy bar and potato chip wrappers were made of paper, but now they are made of plastic—and found prominently in litter streams.

Keep America Beautiful (KAB) released its *2020 National Litter Study,* which provides an estimate of the litter on the ground and in waterways across the United States. Cigarette butts (made with plastic and paper fiber) remain king as the number one item in the litter survey. Plastic films and food-packaging films take the number two and number three spots. When categorizing by material type, items made from plastic topped the list at 38.6 percent, followed by paper at 15.2 percent, metals at 7.9 percent, and glass at 7.2 percent.[21] Plastic litter abounds inland, on the beaches, and in the waterways.

Single-use plastics are a significant drain on the economy. Consider that each local government in the United States pays for the disposal of plastics in their residential trash collection systems and plastic sortation and removal costs in their recycling systems. For example, local governments in California spend about $25 million annually to dispose of single-use plastics, according to a state-sponsored study.[22] In addition, the World Economic Forum estimates that plastic packaging loses 95 percent of its material value after its initial use, at an economic loss to the economy of $80–$120 billion annually.[23]

Although environmental groups have highlighted the problem of single-use plastics so well over recent years, there has been an eerie silence about the remaining plastics produced and utilized. While plastic packaging represents 26 percent of total plastic volume (this total is estimated at 368 million metric tons in 2019 and expected to double over the next twenty years[24]), the remaining three-quarters of global plastics is produced for "durable goods" and other consumer and industrial plastic components.[25] This is the category we hear less about, yet these plastic products also end up in our landfills and litter streams. When did you last recycle the plastic components of your microwave, refrigerator, or leaky faucet when you chose to replace it? Not likely. These plastics are not heading to the recycling streams but to disposal, or worse yet, the oceans.

The point is, regardless of the designation of "single-use" or "durable," plastics impact the climate, human health, and the local environment.

"ADVANCED RECYCLING" AND "CHEMICAL RECYCLING" OF PLASTICS

The typical approach to recycling plastics is known in the recycling industry as "mechanical recycling," where plastics are physically sorted into separate resin categories by hand, assisted by optical sorters and mechanical sortation equipment; baled in separated commodity streams; and shipped to appropriate secondary markets. At the plastic mill, the bales are broken, shredded, chipped, or pelletized, washed, and impurities discarded (usually to a landfill). The final chips or pellets are ready for sale to the plastics injection or mold manufacturing markets. Thus, recycling creates a feedstock material for new products without destroying the integrity of the initial plastic material.

This process requires the material to maintain its essential properties and not be destroyed or manipulated. This mechanical recycling process is delineated through the National Recycling Coalition (NRC) definition of recycling: "Recycling is a series of activities by which material that has reached the end of its current use is processed into material utilized in the production of new products."[26] Therefore, the energy deployed to fully recycle a product into another product is always less than the initial energy inputs to create the initial product.

An alternative nonmechanical process has appeared in recent years: advanced recycling—aka chemical recycling. As the plastics industry uses the two terms interchangeably in its publications, I will call this process *chemical recycling* in this discussion. Unlike mechanical recycling, chemical recycling is an extensive process where plastics are broken down into their original chemical components. The breakdown process leads to an output at the molecular level, destroying the integrity of the original plastic product. Chemical recycling aims to create new industrial fuels, oils, waxes, and other petroleum products, or in some cases, plastic products.[27]

Pyrolysis and gasification are process forms applied to plastics in chemical recycling, and in some situations, solvent treatments are utilized. Examples of chemical recycling include the *dissolution* process to dissolve plastics to extract polymers, the *depolymerization* process using solvents and heat to extract polymers and monomers, and the *conversion* process that uses intense reactor heat to extract a fossil fuel feedstock.[28]

As a career recycler, I have significant objections to calling this *recycling* because the deployed processes destroy the original integrity of the material and thereby violate the definition of recycling. However, the actual material questions the fossil fuel industry must answer to their investors are whether the physical and chemical energy inputs to fuel the process exceed or are less than the energy value of the output feedstock and whether the monetary resource investments have a reasonable return on investment. What is often not asked is whether these chemical processes harm the environment and humans.

A typical 100,000-ton/year chemical recycling plant can cost $300 million in initial capital costs, and the ongoing processes are energy-intensive and rely on external energy inputs.[29] In addition to the direct GHG emissions from the operations, chemical recycling further aggravates climate change by perpetuating the continued extraction of fossil fuels for plastic production.[30]

The plastics industry notes that mechanical and chemical recycling are not competitive industries but complementary approaches with different purposes in recycling plastics.[31] The sales pitch I hear at conferences is that chemical recycling is the savior to the woes of the plastic recycling markets—that chemical recycling can handle the low-end plastics that mechanical recycling cannot. This is a sales pitch we should be wary of, as it comes from the fossil fuel industry that desires to produce more plastics through gas and oil extraction. The poor logic goes that if the low-end plastics have an end market after their initial function is served, then there is no need to worry about banning these products or restricting their sale or use.

A report looking into the nature of chemical recycling identifies thirty-seven plastic chemical recycling facilities proposed since 2000, with only eleven currently operational. None produce plastic-to-plastic, but all produce plastic-to-oils. (One facility claims plastic-to-plastic in a low-volume experimental manner.) Eight of these eleven facilities produce fuels intended to be burned, either as industrial or vehicle fuels.[32] This is very problematic, as plastic-to-fuel facilities create an extreme environmental burden on local communities, much the same as noted in the fracking and cracking facility communities discussed in Chapter 4. In addition, the end fuel products feed the oil and gas industries, contributing to global warming and directly affecting climate change.

There's no getting around the fact that chemical recycling causes harm to the environment and to the climate. Every chemical recycling facility is required to have a permit for air emissions. These air emissions do affect the local communities they are placed in, as well as impacting climate change. In addition, four of the eleven facilities are registered as generators of hazardous waste with the USEPA and their local states.[33] For each ton of plastic processed in a pyrolysis chemical recycling facility, the facility "emits at least 3 tons of CO_2."[34] Plastics-to-fuel facilities "[emit] CO_2 in both production and burning of plastic-derived fuel, which is another fossil fuel."[35] Each ton of CO_2 emission adds to the acceleration of climate change, as well as poor air quality downwind from each of these facilities.

There is a strong environmental justice component to the siting of these facilities: "eight of the plants are located in areas with lower-than-average levels of income, compared to the national average; and five have higher-than-average concentrations of people of color than the rest of the country."[36] Environmental justice concerns affect the citizens in these communities and the workers in these facilities. "Those who will bear the highest of these risks are the nearby residents and workers, many of whom are already impacted by toxic emissions of petrochemical refineries and plastic production facilities."[37]

A technical assessment report of chemical recycling notes that "the technology [is] polluting, carbon-intensive, and riddled with system failures, disqualifying it as a solution to the escalating plastic problem, especially at the scale needed."[38] In another report, the authors traced failed chemical recycling projects globally, including the Thermoselect facility in Germany, which lost more than $500 million, and Interserve in the United Kingdom, which lost $100 million on various chemical recycling projects, with numerous other companies facing bankruptcy.[39]

Despite these documented financial failures worldwide, companies are making significant investments in new chemical recycling plants. ExxonMobil, one of the largest fossil fuel companies in the world, announced an initial investment of $200 million to build out its chemical recycling operations in Beaumont and Baytown, Texas, with an operational start date in 2026. ExxonMobil also announced a partnership with Agilyx and LyondellBasell to construct a chemical recycling plant near Fort Worth, Texas.[40]

Plastics-to-fuel does not solve the plastics overproduction problem; it's simply a trophy show-and-tell project for the fossil fuel industry that lets them justify the production of more plastics. "See, we are solving the plastics crisis," they say as they add more fuel to the fire. Furthermore, plastics-to-fuel is not recycling. Let's call it what it is: material destruction that causes environmental and human harm.

The World Wildlife Fund (WWF) recently published a position paper on chemical recycling implementation principles. Their statement is not an "endorsement of any chemical recycling technologies." Instead, the position paper offers ten principles for the recycling industry to consider regarding the environmental impact of such technologies on human health and nature. One such principle is: "Chemical recycling systems should not transform recyclable material into non-recyclable material."[41]

The Recycled Materials Association (ReMA) (also known as the Institute of Scrap Recycling Industries or ISRI), a recycling industry association representing more than 1,700 companies in the United States and forty countries around the globe, posted a position on chemical recycling that states: "Non-mechanical processes that convert plastics at the end of life into petrochemical products that are fuels or used to make fuels do not meet ISRI's above definition of plastics recycling and thus cannot be properly considered recycling."[42] The National Recycling Coalition, representing 4,500 recycling members in North America, recently adopted a similar policy position, stating, "Non-mechanical processes that convert plastics into petrochemical products that are fuels or used to make fuels, gases, oils, or waxes do not meet NRC's definition of recycling."[43]

The national recycling organizations are united in their concerns about chemical recycling. However, the plastics industry is divided about the new direction of chemical recycling. In May 2021, the Association of Plastics Recyclers (APR) announced a position statement that chemical recycling should only include processes converting plastic-to-plastic resins, departing from the direction of some industry users of plastic feedstock for fuel or energy. In addition, APR's position statement supports design for recyclability, which the mainstream recycling industry has long been pushing.

For those interested in reading more on the harmful effects of chemical recycling, I urge you to read *All Talk and No Recycling: An Investigation of the U.S. "Chemical Recycling" Industry*, a 2020 report authored by Denise Patel, Doun Moon, Neil Tangri, and Monica Wilson.

COGNITIVE DISSONANCE: MISSING THE THREE RS AND LCAS

Let's revisit the human brain and the concept of cognitive dissonance from Chapter 5. According to Festinger, the psychologist who originated the term *cognitive dissonance*, the goal of our brain is to seek consistency with our beliefs or past conceptual understanding. Therefore, new ideas and concepts must pass through this unconscious bias. Humans are subject to this form of nonrational decision-making, yet we must know its dangers and potential pitfalls. We can miss opportunities for change and make poor judgments by ignoring the rational decision tools available to us.

I will reintroduce a concept you learned in grade school and possibly forgot—the three Rs: Reduce, Reuse, Recycle. The basic idea here is that the highest and best use of our resources is to reduce waste first, reuse if possible, and finally consider recycling. Now, don't jump the gun and go straight to recycling. Unfortunately, a cognitive dissonance appears to be creating a bias toward recycling as a solution to all our environmental problems. It's our knee-jerk reaction.

Yet, in this plastics crisis, the focus we need is the first two Rs. We need to look upstream at waste reduction. How can we avoid the use and "need" of so much plastic? How can we reuse the plastic that does exist in front of us before recycling it? Reducing and reusing takes much less energy investment and has lower carbon footprints than recycling, and offers more potential than we are currently recognizing as strategies to mitigate waste. The order of the three Rs is purposeful; it teaches us to work on reducing and reusing first, then recycling, yet our brains jump straight to recycling. Cognitive dissonance offers this unconscious decision, jumping to the conclusion that "recycling can solve the plastics crisis" and telling us "our disposable lifestyle is all right so long as we recycle." It all fits neatly into that unconscious bias of nonrational decision-making. We must combat this cognitive dissonance and allow awareness, new beliefs, and fact-based concepts to push their way past this unconscious bias—because there are rational choices out there to study and act on!

Another cognitive dissonance I wish to highlight is the overreliance on life cycle analyses (LCAs). As a brief explanation, LCAs extract measurement through a series of criteria (e.g., energy, water, material use, etc.) in the various stages of the product's life cycle (e.g., material extraction, transportation, manufacture, packaging, use, end-of-life disposition). The USEPA defines LCA as a "cradle-to-grave" method to evaluate the environmental effects associated with any given industrial activity.[44] This involves a lot of data collection and distills the equation of a product's lifespan down to numbers. The goal of a product LCA is to create a scientific methodology for assessing the environmental impact of a product or service for product comparisons and product improvements.[45]

Although there is significant science behind LCA, it is simply a tool. Like any tool, it has its limits and drawbacks, so problems arise when there is blind adherence to the direction of an LCA. There is a cognitive dissonance involved in this scenario where companies use it as justification to design and produce more plastics. Many LCAs I have seen that appear to "justify" the use of plastics over other materials, such as glass jars or paper takeout containers, have significant *assumptions and data inputs* that are either outdated or not in use today. The cognitive dissonance I speak of is that today's *climate science* adds a new dimension that is *not* factored into the traditional product LCA calculations.

For example, the single-use plastic water bottle has a better environmental LCA score than a recyclable glass bottle, even though that plastic bottle was initially filled several thousand miles away in the Everglades from a bottling plant that is threatening to drain the freshwater supplies of the natural region. The LCA doesn't account for any of that. It likewise does not account for a circular economy where the recyclable glass bottle is manufactured by local recycling programs within fifty miles of its end use. Unfortunately, most LCAs assume past practices of long-distance hauls, not new circular economy practices with reduced transportation models.

Another well-publicized LCA analysis displays film plastic packaging as superior to any recyclable paper packaging, even though its end-of-life destination is the landfill. But, again, this analysis is beholden to the numbers without a complete understanding of the environmental landscape involved with climate change. The answer does not lie in the lifespan analysis but in the search for waste avoidance. Cognitive dissonance thinking leads us down the track of blind allegiance to the results of LCAs without questioning the inputs. There is an adage for any data management tool: Garbage in, garbage out.

A typical product LCA generalizes the pathway traveled from inception to disposal as a common average of environmental impacts. A lot is missing in that picture: supply chain disruptions, fuel supply changes, severe weather impacts, unauthorized changes in engineered specs, and many other variables. A widespread criticism of LCA concerning recycling is that it doesn't factor in product waste reduction and reuse opportunities, thus not offering a fair comparison of material types for each product used. In addition, typical end-of-life options for the product and their usefulness and toxic dangers are not measured through LCAs, again not providing a fair comparison of material types. Simon Hann wrote on these issues in an excellent report that explores the entitlement of plastics in LCAs, titled *Can Life Cycle Assessment Rise to the Challenge?*[46]

The alternative to the cognitive dissonance of LCA tracking is to move toward rational decision-making processes. Question the assumptions behind the data tools and input necessary data that fits circular economic thinking. For example, transporting single-use drink bottles halfway around the world is old-school thinking. Developing local jobs in a remanufacturing hub utilizing local recyclables and reusables to fuel a local circular economy is the new future. But is it realistic?

New data models need to be developed and explored. None of the LCA studies I have seen factors in the destructive forces of the fossil fuel industry in creating plastic products. Most of these plastic LCAs begin data collection at the creation of plastics from oil or gas, yet do not factor in the environmental attributes of the feedstocks of oil and gas from extraction to the creation of the plastic. Yet, the fact remains: Plastics are fossil-fuel-generated products. This is the underpinning of the plastics crisis and a significant source of environmental harm that needs to be accounted for in LCAs. The climate impact facts are there. A decision made without looking at those facts is sorely misguided.

MOVING TOWARD ZERO WASTE

The basic concept behind Reduce, Reuse, Recycle is a waste management hierarchy that lays out actions an individual can take to divert waste from the landfill or the incinerator. These are clean green activities, often viewed as the first entry point into the home for more aggressive environmental actions.

Zero waste is not a waste management strategy, nor is it a more advanced form of recycling. Rather, zero waste is a sustainable resource management strategy, defining our discards as resources deserving a second life. Instead of the trash can, material discards are prepared for additional uses. Zero waste is a different frame of thinking about Earth's resources, embracing the Waste Not, Want Not philosophy of the past and moving forward through greener product redesign to ensure products and packaging impact the environment and humans to a lesser degree.

The Zero Waste Hierarchy places resource management strategies in the order of the highest and best use.[47] The starting point with zero waste is not recycling but rather a new set of three Rs: Refuse, Rethink, Redesign. Consumers and producers can start by *refusing* what they don't need, *rethinking* their overconsumption of plastics, and changing how they produce and consume by *redesigning* products and packaging to reduce environmental impact and waste.

The second series of actions in the Zero Waste Hierarchy is "reduce, reuse, recycle, and compost." We all learned these three Rs, yet we often forget that waste reduction comes first, then reusing existing products before considering recycling or composting.

Notice that final disposal is a last-resort consideration. Both corporations and individual consumers have a social responsibility to keep this in mind. Extended producer responsibility (EPR) is often a legislative response toward requiring manufacturers to take responsibility for the end-of-life management of their consumer products. This policy encourages green product design. As individual consumers, we should make full efforts to extend the life of our products before disposal. This is contrary to the plastics disposable lifestyle pushed by the fossil fuel plastics industry—and that's exactly the point.

Deploying the Zero Waste Hierarchy on the production of plastic products and packaging would imply the development of more durable, longer-lasting products, the elimination of single-use disposal products and packaging, research and development into safe plastic substitutes, and the introduction of reusable beverage containers. Rethinking the common use of plastics through a zero waste lens would be a starting point in addressing the global plastics pollution problem.

Refuse, Rethink, Redesign, Reduce, Reuse, Recycle, Compost . . . then Dispose.

Artwork by Pam Longobardi + Drifters Project

A DISTANT MIRROR, 2014

Ocean and urban plastics from Hawaii, Costa Rica, Italy, Greece, Alaska, California, and Atlanta; steel pins; and silicone, 110" x 68" x 4". *A Distant Mirror* is made from both oceanic and urban plastic. I intervene in the journey of escaped plastic I encounter on the streets of Atlanta, preemptively capturing it and interrupting its flow toward the sea. I use the form of the mirror as a conceptual model to see the reflection of ourselves in the vagrant plastic objects I recover. The piece doesn't provide an external reflection but is intended to provoke an internal reflection.

—PAM LONGOBARDI

Chapter 7

SPACESHIP EARTH

"We are not going to be able to operate our
Spaceship Earth successfully nor for much longer
unless we see it as a whole spaceship and our fate
as common. It has to be everybody or nobody."

—R. BUCKMINSTER FULLER[1]

The 1970s brought about environmental awareness throughout the nation, and the world, with the kickstart of the first Earth Day on April 22, 1970. As a teenager at that time, I watched Lake Erie die out from its natural basin to become a cesspool of industrial chemicals, and I wondered why "we" collectively as a society allowed Earth to be so polluted. Why did we treat Mother Earth—our home, our ancestors' home, and home to our future generations—so poorly? Was it because we didn't read the "operating manual," as Buckminster Fuller once noted?

My grandfather's farm was a consoling place for me to ponder. Sometimes I wondered alone as my parents chatted in the house; sometimes, I wondered with the wise console of my grandfather as we walked around the farm. He didn't offer solutions to the world's problems but taught me the workings of the family farm. Living in harmony with nature was the common thread of his words of wisdom. He reduced his tilling depth and timing to reduce soil erosion (this was before "no-till" became a demonstrated farming technique). He utilized cover crops to increase moisture retention. He maintained an annual crop rotation to ensure soil quality. He added only natural nitrogen from the chickens, if you catch my drift.

Three lifelong lessons I picked up from my grandfather's farming have benefited me throughout my life and allowed me to tread lightly on this earth. The first is the concept of harmony with nature. Nature is a friend, not an enemy to fight and conquer. Grandfather attempted to work with the natural elements rather than against them. Of course, he wasn't a purist, and occasionally, he had to utilize herbicides or pesticides (I cringe), but he targeted them rather than over-spraying. In addition, he respected the animals from the nearby woods and learned to create buffers between farmland and natural areas.

The second concept I learned was to watch for signs of changing weather patterns and base your future activities on the upcoming weather. For example, he often said, "Don't plan a picnic on a predicted rainy day." That's a fundamental concept, of course, but many people ignore current weather forecasts and become frustrated when the weather does not fit their plans. Grandfather knew the rains, wind, and snow were necessary elements of life, and we must fit into the seasons rather than trying to force the daily weather to accommodate our busy schedules. This is extremely important as climate change begins to play havoc with our "normal" weather patterns.

The third life concept I cherish from my grandfather is the power of the seed. Once planted and nurtured, the tiny seed can grow into a plant, fruit, vegetable, or even a mighty oak tree. If planted in rocky or infertile soil, it will fail. If not watered, it will not grow. If not nurtured, it will not succeed. But with the proper care, usually through a healthy Mother Earth, the mighty seed will grow to its potential. To a small family farmer, every day is Earth Day.

Every grade school science class brings us back to the wonders and powers of the seeds planted in soil and cared for, bringing forth that unique plant or flower. Environmental leader Dr. Vandana Shiva's essay collection *Sacred Seed* displays the sacredness of seed-growing rituals and mysteries over the centuries, and I recommend it as a worthy read.[2] The seed's life bears witness to Earth's regenerative capacities to regrow and renew its spirit. It is our duty to tend to the same.

PLANETARY FRAGILITY

Upon returning from space on Apollo 11, Neil Armstrong stated: "It suddenly struck me that that tiny pea, pretty and blue, was Earth. I put up my thumb and shut one eye, and my thumb blotted out the planet Earth. I didn't feel like a giant. I felt very, very small."[3] William Shatner also commented on his view of Earth after his 2021 space trip: "I wept for the Earth because I realized it's dying," Shatner said. "I dedicated my book, *Boldly Go*, to my great-grandchild, who's three now—coming three—and in the dedication, say it's them, those youngsters, who are going to reap what we have sown in terms of the destruction of the Earth."[4] He described his initial reaction as grief for a dying planet and a dedication toward action for future generations, and he's not alone in these realizations.

I met Richard Garriott de Cayeux at an Austin Earth Day celebration, where he was the guest speaker, displaying his presentation slides of his trip to space and his favorite photo of Mother Earth. My honor was introducing him on stage; however, the additional honor later was to enjoy having lunch with him and talking about those special twelve days in space. He too shared about the fragility of Earth as he saw it from a great distance. The vantage point from the space capsule gave him a renewed interest in speaking out to protect Earth for future generations. After returning to Earth, he changed his lifestyle to a higher environmental standard, became more involved in local environmental activities, and later was inducted into Austin's Environmental Hall of Fame.

Astronauts returning from Earth often have what is known as the *overview effect*, a term offered by space philosopher and author Frank White that describes a psychological phenomenon many astronauts experience as a result of looking down on Earth from space in which their view of the planet and life shift. As he describes, "I would see it from a distance. And I would see it as a unified whole. There are no borders or boundaries. All of these things would become knowledge. Living on the surface, we find it difficult to grasp, or mentally grasp philosophically."[5] Alongside this overview effect, astronauts also carry a sense that Earth is a singular home in our galaxy, with a fragile, thin atmosphere against the elements of space—and against the destructive tendencies of humans.

The concept of this planet being "fragile" and needing care and repair from human destruction is well displayed in literature and documentaries. Many of the images we see in early science classes offer the opportunity to care for our common home. Yet the climate crisis brings us back to the reality of an emergency needing immediate attention.

That emergency is best described in the Planetary Boundaries model presented by Johan Rockström and colleagues, all internationally renowned scientists. Their goal was to define a "safe operating space for humanity." Their model identifies nine planetary processes that are key to regulating the stability of Earth, and sets a "boundary" for each not to be surpassed. Crossing any boundary creates a worldwide risk of "irreversible environmental change."[6] Two challenges the scientific community found with this model are quantifying each boundary and understanding the effects of crossing that boundary.

The nine boundaries, now expanded to ten, include: climate change, ozone depletion, atmospheric aerosol load, ocean acidity, freshwater consumption, chemical pollution, agricultural land use, biodiversity loss, nitrogen flow, and phosphorus flow. This is important to this book's discussion because the scientific community has universally agreed that three of these boundaries have already been far surpassed: biodiversity loss, nitrogen flow, and climate change.[7]

In continuing his research on these planetary boundaries, Johan Rockström notes that they found many interactions between planetary boundaries and that the two core boundaries of climate change and biosphere integrity contributed to more than 50 percent of the strength of all the planetary boundaries' interactions in a network combined. Thus, the risks of destabilizing these two boundaries are very high and can "amplify human impacts on Earth's systems and thereby shrink the safe operating space for our children and grandchildren."[8]

There is a glimmer of hope in Rockström's report: "If we reduce our pressure on one planetary boundary, this will in many cases also lessen the pressure on other planetary boundaries. Sustainable solutions amplify their effects—this can be a real win-win."[9] As Earth is fragile in many ways, the Planetary Boundary model is one scientific approach that helps us look at the problem, see where the injuries are, and attempt to offer solutions. Again, that leads us to the climate change issue of rising global temperatures resulting from CO_2 releases caused directly by fossil fuel burning, with plastics as a contributor.

The Planetary Boundaries report sets the boundary for climate change at "350 parts per million by volume CO_2 concentrations" and identifies its current level nearing 400 parts per million[10] (now more than 427 ppm today and rising). By Rockström's report, if climate change is better controlled, it may positively affect the other planetary boundary concerns. Most scientists are directing attention toward reducing the burning of fossil fuels as the most direct means of reducing CO_2 releases, thereby reducing global temperatures. If the mission is to eliminate the burning of fossil fuels, it's necessary to change our transportation systems from burning fossil fuels to using renewable energy. Yet that is not the only change required. Other sources of CO_2 releases include the creation, use, and disposal of plastics. *Throughout the entire life cycle of synthetic plastics, CO_2 and CH_4 releases impact the atmosphere.*

The planetary boundary has been breached—it's time to repair that breach. The repair cannot be partial and incomplete; we must seal off all unnecessary and unintended releases of greenhouse gases, and we must begin this effort immediately. To move toward a sustainable lifestyle within what may be called a circular economy, the release of gases must be done with purpose, and all gases must be accounted for. Unmitigated released gases into our atmosphere are unacceptable, given the planetary emergency we now live in.

OPERATIONAL INEFFICIENCY

R. Buckminster Fuller once summed up the concept of inefficiency in business terms: "Any waste as an output from a business is an operational inefficiency." I mentioned this earlier in Chapter 1 but feel the need to revisit this topic. The concept is simple: Run the business with as few inefficiencies as possible. Streamline through conservation methods.

We are now at the point where we need to look at the operational efficiency of human impacts on Earth and to regard any output onto Earth's systems that was not intended for a specific activity as an unacceptable operational inefficiency that must be eliminated. That includes unmitigated gas releases into our global atmosphere and plastic disposal into our international waterways.

A simple example of this approach is the incandescent bulb. This type of bulb has supplied light for a century. Yet the bulb could be more efficient in using electricity as it provides a form of unintended heat pollution that is not serving its specific purpose of providing light. Replacing the incandescent bulb with an LED bulb eliminates heat pollution and conserves electrical energy, thus solving two operational efficiency issues. (Unfortunately, the LED bulb conversion did not resolve the recyclability issue.)

Eliminating the use of plastics and providing for the substitution of other materials is an approach toward resolving the inherent "operational inefficiency" of plastics. Given the human, animal, and Earth impacts of plastics noted in previous chapters and the global emergency of CO_2 releases contributed by plastics, eliminating plastics will offer a pathway toward operational efficiencies and planetary recovery. However, the path requires *careful consideration of the replacements chosen*, as certain materials and processes may cause more harm to the environment and humans than intended. We must heed this caution, but that certainly is not a stop sign but rather a challenge for the innovators!

OUR HOME, SPACESHIP EARTH

Many years ago, R. Buckminster Fuller famously noted, "Now there is one outstanding important fact regarding Spaceship Earth, and that is that no instruction book came with it."[11] An interesting thought: I have an operating manual for my computer, my printer, my washer and dryer, and all my appliances. Geologists can analyze rock formations, and paleontologists can offer a history of Earth based on fossils. But there is no operating manual for Planet Earth.

Climatologists (scientists who study Earth's climate) have demonstrated a *direct relationship* between CO_2 emissions and the average global temperature increase. Using global climate models and historical climate data, researchers have demonstrated a *"linear relationship between total cumulative emissions and global temperature change."*[12] At the same time, it has long been established that CO_2 emissions are the direct product of the burning of fossil fuels, from coal, oil, gas, and diesel. Energy is the intended result, but burning fossil fuels releases carbon dioxide as an unintended side effect of the chemical reaction. Since the beginning of the Industrial Age, the global average CO_2 concentration has risen from below 200 ppm to more than 400 ppm, increasing global temperatures in that time span by 1.5° Celsius.[13] Climatologists observe that the CO_2 concentrations are accumulating faster, resulting in accelerated global warming. To put that in nonscientific terms, *burning oil, gas, and coal, creating carbon dioxide emissions, is warming Earth faster than ever recorded in human history.*

The Paris Agreement, signed by 196 nation-states in December 2015, set an ambitious goal of limiting global warming to well below 2.0°C, with a global preference toward 1.5°C, compared to pre-industrial levels.[14] Yet most of these industrialized nations are directly authorizing gas and oil permits to produce about 50 percent more fossil fuels by 2030 than would be consistent with limiting warming to 2°C and 120 percent more than would be consistent with limiting warming to 1.5°C.[15]

Stop for a moment and digest that. The world leaders met in Paris, hashed out an agreement among 196 nation-states to address the climate crisis, and then went back home and continued business as usual, including issuing oil and gas permits that directly violate those international agreements. In addition, the G20 governments, as of November 2020, have committed $233 billion to "activities that support fossil fuel production and consumption," according to the report issued by The Production Gap.[16] The report shows that the U.S. government has dedicated more than $100 billion in federal public funds to support the fossil fuel industry. This spending contradicts the commitments made within the Paris Agreement and charts a reckless course toward the tipping point—the point of no return—on the climate crisis.

On August 9, 2021, the Intergovernmental Panel on Climate Change (IPCC) held a press conference to release the first installment of the IPCC's Sixth Assessment Report (AR6). It is important to understand the key points of the announcement:

- Over the next 20 years, there are higher chances of exceeding the global warming level of 1.5°C.

- Climate scientists call for the immediate and widespread limiting of greenhouse gas emissions.

- A clear declaration that carbon dioxide (CO_2) is the primary factor driving climate change.[17]

This report includes disturbing data and information that should set off alarm bells worldwide. It's a wake-up call to all who have been complacent and felt that action could be delayed. But however gloomy the report may seem, it also spells out a survival route that requires human effort: "Stabilizing the climate will require strong, rapid, and sustained reductions in greenhouse gas emissions, and reaching net zero CO_2 emissions," as well as reducing other pollutants such as methane.[18]

Astronomer Carl Sagan warned us, "We have a choice: We can enhance life and come to know the universe that made us, or we can squander our 15-billion-year heritage in meaningless self-destruction. What happens next depends on what we do, here and now, with our intelligence and our knowledge of the cosmos."[19]

"Meaningless self-destruction" may seem like harsh words, but when looking through the prism of a planetary emergency, as we should, this is a wake-up call. Finally, a decision must be made: Shall we wean ourselves from fossil fuels (and plastics made from fossil fuels), or shall we sacrifice the planet's habitat, our spaceship Earth, all in the name of money and convenience? *Who makes that decision? Who has the right to destroy our planetary future?*

ECOCIDE: HOLDING POLLUTERS RESPONSIBLE

If a company or nation creates a massive environmental form of destruction on a global scale, shouldn't that be considered an international crime similar to genocide? The concept dates back to a Swedish proposal to the 1972 U.N. Conference on the Human Environment.[20] Today, international lawyers are attempting to define *ecocide* for the world criminal courts.

A legal definition for the term *ecocide* has been proposed to the Rome Statute of the International Criminal Court (ICC), authored by the Stop Ecocide Foundation, a Netherlands-based coalition, and the Independent Expert Panel for the Legal Definition of Ecocide. In June 2021, a proposed definition was unveiled for international lawyers to work with and debate the beginning of a long road toward adopting a final report. The draft definition is as follows:

For the purpose of this Statute, "ecocide" means unlawful or wanton acts committed with knowledge that there is a substantial likelihood of severe and either widespread or long-term damage to the environment being caused by those acts.

For the purpose of paragraph 1:

1. "Wanton" means with reckless disregard for damage which would be clearly excessive in relation to the social and economic benefits anticipated;

2. "Severe" means damage which involves very serious adverse changes, disruption or harm to any element of

the environment, including grave impacts on human life or natural, cultural or economic resources;

3. "Widespread" means damage which extends beyond a limited geographic area, crosses state boundaries, or is suffered by an entire ecosystem or species or a large number of human beings;

4. "Long-term" means damage which is irreversible or which cannot be redressed through natural recovery within a reasonable period of time;

5. "Environment" means Earth, its biosphere, cryosphere, lithosphere, hydrosphere and atmosphere, as well as outer space.[21]

To be clear, this definition still needs to be cemented in international law. First, it must go through an ICC legal review process, then on to the General Body of the United Nations, where a two-thirds vote of the nations is required for adoption. Although this is a draft definition, it's a great start in the right direction.

Consider the possibilities: *a group of fossil fuel companies being prosecuted for harm to our Planet Earth—under the definition of ecocide!*

Not far from this concept is the court's action against the oil company Shell. An NPR news story states, "A Dutch court ruled that the company [Royal Dutch Shell] must reduce its greenhouse gas emissions 45% by 2030, based on 2019 levels."[22] Seven environmental groups filed the lawsuit against Shell, with 17,000 Dutch citizens as co-plaintiffs.

The court ruling was far-reaching as the greenhouse gas emissions restrictions also apply to Shell's product customers.[23] Furthermore, the court ruled "that Shell's activities constituted a threat to the '*right to life*' and '*undisturbed family life,*' as set out in the European Convention on Human Rights."[24] Marit Maij, executive director of ActionAid Netherlands, one of the organizations that brought the case against Shell, was quoted as saying: "Big polluters beware. This landmark ruling now sets a precedent that corporations can be held liable for causing runaway climate change and forced to cut emissions in line with global climate goals."[25]

Legal change is also coming on other fronts. On May 12, 2021, plastic was declared "toxic" under Canada's primary environmental law, the Canadian Environmental Protection Act (CEPA). This decision was driven by a study commissioned by Environment and Climate Change Canada (ECCC) that found that 3.3 million tons of plastic are discarded in Canada each year, and less than 10 percent recycled.[26] "Adding plastic manufactured items to Schedule 1 of (CEPA) will help us move forward on our comprehensive plan to keep plastics in the economy and out of the environment," said Moira Kelly, press secretary to Environment Minister Jonathan Wilkinson. The goal then would be "to implement our proposed ban of certain harmful single-use plastics, make producers responsible for their plastic waste, and introduce recycled content standards."[27]

David Arkush, the climate director at the advocacy group Public Citizen, and Donald Braman, an associate professor at George Washington University Law School, began hosting a series of panels at prominent law schools in 2024 to discuss the concept that fossil fuel companies should be charged with homicide. Braman said, "We're talking about the idea that these corporations had a deep and detailed understanding of what they were doing; they really tried to hide that from the world as best as they could, and they were very effective at driving doubt and delay into the market, into our democracy."[28]

Karen Wirsig, program manager for Environmental Defense, stated, "We need to reduce the amount of plastic that gets put on the market and, therefore, into the environment. We need to find alternatives to plastics in many cases."[29] I could not have said it better!

URGENCY FOR ACTION

The urgency is readily felt when one reads the August 2021 IPCC report. The IPCC press release notes that the report estimated the probability of crossing the global warming level of 1.5°C, finding that we will likely cross that limit unless rapid and widespread reductions in greenhouse gas emissions are immediately taken into effect. In addition, the report also notes that climate changes will increase across the globe: "increasing heat waves, longer warm seasons, and shorter cold seasons" at 1.5°C and climate extremes that "would more often reach critical tolerance thresholds for agriculture and health" at 2°C.[30]

Another alarming study by an international group of researchers discovered that CO_2 and other GHG emissions are "shrinking" the stratosphere. Co-author Juan Antonio Anel stated, "We discovered that the stratosphere has been contracting by more than 100 meters [328 feet] per decade since 1980, and we have proved that it's due to greenhouse gases." At this rate, the stratosphere "could lose 4% of its vertical extension (1.3 km) [0.8 miles] from 1980 to 2080"—one more example showing humans' damaging effects on "the fragile balance of our planet."[31]

Down on Earth, we just experienced the warmest decade ever recorded in weather history, 2010–2019. "Since 1980, each successive decade has been warmer than any preceding decade since 1850," according to the World Meteorological Organization (WMO).[32] The current decade, 2020–2029, is on pace to be the hottest globally, outpacing the previous decade. The WMO report displays not only the effects on the weather but also the human toil. "Worldwide, some 6.7 million people were displaced from their homes due to natural hazards—particularly storms and floods, such as the many devastating cyclones and flooding in Iran, the Philippines, and Ethiopia."[33] WMO credits these displacement figures as *forced climate migration.*

The writing on the wall is clear: Change was needed in the last decade, last year, last month, but the second-best time to act is now—before it really is too late. So what can be done? Where do we go? How can we help?

Hope for the future requires looking through a different set of lenses. If we remain rooted in past views and past ways, nothing changes. Various future views are necessary to generate different results from your actions. For example, consider the path of a contrarian. We must "unfollow" the standard pathway and move forward with a new decision pathway. Now is the time to look and act boldly.

"We have been told so many times that there's no point in doing this, that we won't have an impact anyway, that we can't make a difference. I think we have proven that to be wrong by now. Throughout history, the most important changes in society have come from the bottom up, from the grassroots."

—GRETA THUNBERG[34]

Artwork by Pam Longobardi + Drifters Project

BOUNTY, PILFERED (FOR BP), 2014

Over one thousand pieces of vagrant ocean and urban plastic recovered from Alaska, Greece, Hawaii, Costa Rica, Atlanta, and the Gulf of Mexico; steel armature and wire; drift nets and floats from the North Pacific Gyre collected in Hawaii and Alaska, 62" x 78" x 150". The title represents the ubiquitous waste associated with disposable plastic as a "squandered horn of plenty," while the black color of plastic points to its origin in oil. The title, *Bounty, Pilfered*, also shadows the initials of BP, the oil conglomerate responsible for the Deepwater Horizon spill that has forever altered the bountiful life of the Gulf of Mexico.

—PAM LONGOBARDI

Chapter 8

A CONTRARIAN VIEWPOINT

"People should think things out fresh and
not just accept conventional terms and the
conventional way of doing things."

—R. BUCKMINSTER FULLER[1]

I had two childhood heroes who shaped my personality during my formative years and drove me to make a difference in this world. First, as you've previously read, my grandfather taught me the farmer's ethic of do no harm to Earth. My second hero was Neil Armstrong, the astronaut who showed me that if a human can walk on the moon, anything is possible when you set your goals with intention and determination.

As Armstrong was preparing for his Apollo spaceflight, there was a nationwide children's contest to write an essay about returning from the moon. The competition was to write a speech that perhaps Neil Armstrong would deliver at a press conference on the ship that rescued him from the splashdown in the cold Pacific waters. Of the thousands of entries submitted, I won that contest! Of course, there was no guarantee that Neil would read the speech I wrote, but I waited for the day of his safe return, only to see him swept away into isolation and confinement before he could deliver any speech for fear of an unknown moon virus or bacteria. Since those days, we now have less fear of such issues, but Armstrong did not issue a press conference as anticipated.

Several weeks later, much to my surprise, I received a phone call from my hero Neil Armstrong, who thanked me for the written speech and the good wishes. Although only a three-minute call, it left a lasting impression on me. I remember thinking, *If Neil could walk on the moon, I can achieve my dreams.*

Why does that story matter so much? Because at that time in my life, I was a heavy stutterer. I couldn't string a complete sentence together with my stuttering. I stuttered on that call with Neil, but he was patient and listened to me. He also told me I could be who I wanted to be, regardless of any speech impediment. *If Neil can walk on the moon, I can learn to talk and achieve much more!* After that call, I was more dedicated to my speech lessons and learning to overcome the stuttering that had burdened me for so many years.

As my career blossomed many years later, I became the district director of the Auglaize County Solid Waste District in Ohio. I assisted in developing curbside and drop-off recycling programs and managed the county recycling processing facility. It just so happens that Neil Armstrong grew up and learned to pilot an airplane in Auglaize County. His first job was in a pharmacy in Wapakoneta, and my office was in the Wapakoneta courthouse. He flew planes at the New Knoxville airport, near where we set up a recycling drop-off in the village of New Knoxville many years later. Although I never crossed paths with Neil Armstrong in Wapak, it was significant to me that I worked in the county where he grew up. We established the recycling program motto: "If Neil can walk on the moon—you can Recycle!" It was a successful recycling program, with a significant majority of residents participating in the first week of the county-wide roll-out.

THE CONTRARIAN PATHWAY

Throughout my fruitful career, I have learned many lessons from many good people. The one lesson that keeps coming back like a boomerang is that of solution-building through community input. Whenever I engaged in community solution brainstorming, in Wapakoneta and many other communities I served in, I often found that the answers that surfaced were surprising, unexpected, and offered new and innovative pathways toward resolving the environmental concerns of the day. Solution-building through community input sometimes creates a bond throughout communities that are most affected by the adverse outcomes of the environmental degradation we face each day. Unfortunately, we often ignore those voices, pretending that the "experts," "officials," and "people in power" will solve these environmental problems.

A quote credited to Einstein seems to apply: "We cannot solve our problems with the same thinking we used when we created them." Likewise, solution-building cannot come from the same thinking processes that created the problems in the first place. If we take climate change seriously and want to change the current climate crisis path, we must diverge from the pathway that led us up this steep and life-threatening cliff. A different pathway is required to generate different results.

Consider the contrarian. A *contrarian* is "a person who takes a contrary position or attitude."[2] Steven B. Sample's book *The Contrarian's Guide to Leadership* offers excellent advice that should be taught in every management class. One of the first things I did in management was ditch the *One Minute Manager* approach (apologies to Kenneth Blanchard) and embrace the quote from Sample's book that says we should "listen first, talk later; and when you do listen, do so artfully."[3] Taking a leadership position requires proactive listening. After listening and hearing all the stories (not all sides, but people's personal and professional stories), you move forward with "critical thinking and discernment."[4]

Both in management and other arenas, we must engage in a contrarian pathway to resolve the climate crisis. We must unfollow the rules of the twentieth century and make new rules for the twenty-first century—for we are, after all, already a quarter of the way into it! We must create a significant paradigm shift (to use a tired old term), and doing so requires a contrarian viewpoint. The old viewpoint of "repairing or fixing the problem" leads nowhere. In contrast, an action-oriented person will seek out the *root cause* and work toward community-based solutions to eradicate that root cause.

In my long career, I, and the communities I have served, have greatly benefited from taking the contrarian position on issues that seemed doomed to perpetual conflicts and dualistic competitions and forging new pathways out of the gridlock. Our society is deeply polarized by partisan politics and sharply divided viewpoints on climate change. Steven B. Sample states that "most people are binary and instant in their judgments; that is, they immediately categorize things as good or bad, true or false, black or white, friend or foe."[5] We commonly establish our debates on these binary dualistic divisions with the thought pattern that all opinions neatly fall within these two categories, if not at the extremes, at least along a spectrum between these two extremes. Binary thinking not only pits us against one another, it also prescribes a strictly linear movement of thought and action.

"In the binary world, there are start dates and finish dates. Things happen sequentially in a linear, orderly fashion," notes Daniel Priestley.[6] He continues on to say that we can move forward from the constraints of binary thinking through an alternative: "Directional thinking is suitable in the domain of uncertainty. It's the thinking required to keep the many moving parts of a business moving 'roughly' towards a desirable outcome."[7] To make decisions in a transformational time, it may be necessary to move even beyond the "directional pathway," stepping into a "contrarian pathway" and taking positions and accepting possibilities contrary to the norm. So, again, we need to unfollow the standard and move forward with a new decision pathway.

THE FOUR DECISION PATHWAYS

This conversation brings me back to my grandfather's lesson on decision-making. As told in Chapter 1, he taught me that when confronted with a contentious problem too big to resolve with simplistic answers, there are four approaches to a resolution. Visualize again the four rectangular cells; the left and right cells represent the common dualistic approach: your way versus my way. This is the binary thinking pattern that generally does not resolve complex societal problems.

The lower cell represents the compromise position. Compromise is the "fail-safe" position when all negotiations and statesmanship fail. I give in some, you give in some, and we compromise somewhere in the middle. Most congressional acts have been adopted through this compromise "sausage-making" process. The recent infrastructure bills in Congress passed this way, yet they failed to address climate change significantly—precisely because of the "need" to compromise. Some statesmen praise the art of compromise; my grandfather said not to stop at this easy pathway. Why? Because compromise brings dissatisfaction almost immediately after the agreement is made, with distrust boiling again. Compromises can be reversed, especially in the world of politics.

He advised taking the long, rugged path toward the top cell of the diagram: *the innovation pathway.* When looking for innovation, a solution that can satisfy all parties involved can appear and create a *trust* that will offer stronger relationships. However, this process of innovation and creating new thoughts can be slower and require patience, listening, deep conversation, and careful deliberation and exploration of the issue at hand. That is a tall order, so compromise is the fast track most often taken when speed is more important than quality. So, before we move on to further thoughts on innovation, let's pause to consider the actual costs of the compromise position.

THE REJECTION OF COMPROMISE

The original draft of the U.S. Declaration of Independence, initially written by Thomas Jefferson, contained a passage blaming King George for the slave trade in the states and condemning the practice of slavery. As presented to the Continental Congress, the draft and its contents were debated. As a result, there was some editing done before it reached its final form for signature—without the anti-slavery clause due to compromise. Jefferson noted in his autobiography that "the clause . . . reprobating the enslaving the inhabitants of Africa, was struck out in compliance to South Carolina and Georgia, who had never attempted to restrain the importation of slaves, and who on the contrary still wished to continue it." As a contrarian, John Adams disavowed slavery completely and spoke out against the practice as the Declaration was being debated. The slavery clause was struck out of the document due to a *compromise* offered to two states to gain their vote for an independent United States. As history will show, this eventually led to the bloody conflict of the American Civil War fought over slavery, an issue left to fester through the compromised edits of the Declaration of Independence.

I offer an alternative scenario—if we could reimagine history. What if that *compromise* had not been acceptable to the writer Thomas Jefferson? What if a coalition of eleven states (not thirteen) had signed the Declaration of Independence with the clause forbidding slavery in the newly formed United States? Perhaps the newly birthed country would have been smaller, with two fewer states, yet slavery would have been outlawed in 1776. Think about the potential massive effects of *not compromising*! We can't rewrite history any more than we can predict how many states would have been brave enough to move ahead or what the outcome of the Revolutionary War would have been minus South Carolina and Georgia. However, we must put on the table the concept of *not compromising* in respect to the human value of freedom from slavery.

Henry Clay, the Speaker of the House of Representatives, has been credited with writing and gaining the votes of the Missouri Compromise of 1820, which granted national legalization of slavery in Southern slave states, preserving a "balance of power" between "free" states and "slave" states. The Missouri Compromise admitted Missouri and Maine to the Union as slave and free states respectively, banning slavery everywhere else north of the 36°30' latitude line while allowing slavery everywhere south.[8] This congressional agreement, through *compromise*, was challenged forty years later in the American Civil War (1861–1865), proving that this *compromise* did not resolve the issue of concern. And ironically, to this day, we call Henry Clay "the Great Compromiser." Many would not call that a Great Compromise.

A more recent example of compromise illustrates the point of potentially acting before a situation reaches a state of emergency. In January 2023, Congress was presented with the usual action to approve the national debt limit. The national debt is the grand total of money the U.S. government owes, either to investors or to itself. The debt ceiling is approved periodically by an act of Congress and must be approved to pay the government bills that Congress previously approved.

Due to the extreme divides in Congress, an agreement to pass the debt seemed unlikely. Although Congress approved the debt ceiling ninety-two times in the past, this time the various sides were staking out their positions, and a divide was inevitable, or so it seemed.[9] It was noted in several news articles that the president waited ninety-seven days to speak directly to the Speaker of the House, as he believed there was no point in rushing into extended talks, "given that no important agreements in Wahington are made until a deadline is looming with catastrophic consequences if the two sides do not come together."[10] That referenced catastrophe would be a default on the U.S. debt by June 7th and the potential ensuing stock market crash.

A "compromise" of historic proportions was reached on June 1st. The House of Representatives and the Senate passed the Fiscal Responsibility Act that suspended the national debt ceiling until 2025. In addition, within the approved bill are significant budget caps and restrictions on spending in the following two budget years, including a rollback of certain preapproved expenditures. But most notably, the bill includes fast-track approval for the highly controversial Mountain Valley Pipeline. For years, communities in Virginia and West Virginia have fought to stop a pipeline that would expose their families and children to a host of dangerous and cancer-causing chemicals. This bill fast tracks the approval of these nat-ural gas lines, counter to the climate change abatement goals of the U.S. government. This is a massive step backward in climate change abatement efforts—all in the name of *compromise*. As Senator Jeff Merkley (D-OR) stated: "We keep greenlighting new fossil fuel projects while we are essen-tially already at the carbon cap, very close to it. . . . And America has burned most of that carbon," even while we preach climate consciousness."[11]

And here's the kicker: Most in Congress are quite unhappy with this "compromise." Yes, it passed, due to the looming deadline to avert the fiscal crisis, yet it exacerbates the climate crisis even more! Senator Merkley said: "The key here is that from the beginning, this was a fail-ure of imagination."[12] That is what all compromises come down to: a failure to imagine a win-win solution, so both parties are forced to set-tle for a lose-lose solution.

What could have been a more imaginative win-win solution here? In November 2022, Janet Yellen, the treasury secretary, suggested that

Democrats in Congress use their remaining time in control of both houses to lift the debt limit beyond the 2024 elections. "Any way that Congress can find to get it done, I'm all for," *The New York Times* notes Yellen saying.[13] After such a fiscal maneuver, Congress could then deal with budgetary cutbacks through the traditional annual budget approval process without the fiscal debt crisis looming overhead. Cooler heads would have prevailed, possibly with more innovative programming ideas. Unfortunately, this scenario is simply a what-if to muse on in the aftermath of a disastrous compromise.

There are many other examples of the poor use of compromise, including human rights violations, war crimes, and environmental disaster cleanup settlements—too many to revisit here, so let us return to this book's purpose. To get straight to the point, *the health of this planet and its inevitable damage through climate change is an existential threat to human health and life itself. As such, compromise is not acceptable. Any pathway toward addressing climate change should not involve compromise—the short-selling of our children's planet. Instead, we must be solid and insistent on taking the innovative pathways that will benefit all societies.*

Building Bridges: A Metaphor

How do we physically build bridges? Bridges are built from both land ends simultaneously until they reach the middle, where they meet and merge—if engineered correctly. Likewise, building bridges between people through challenging issues does not begin in the middle, through compromise. Instead, we listen to both sides with empathy, ask open-ended questions through inquiry, share our stories, and design a future together. We build the bridge from both ends, until we eventually meet in the middle. It's a process that takes time, effort, and intentionality.

RETHINK—THE LARGER PICTURE

Climate change is a multi-issue, complicated web of chain reactions resulting in uncontrolled and ever-increasing warming of Earth's atmosphere. The consequences are growing exponentially as we waste time through our debates. Unfortunately, arguments and compromises slow us from finding and implementing the innovative solutions that we seek.

It's time to explore more innovative platforms and solutions that might help us resolve our stalemate. That does not imply high technology but rather a collaborative, innovative, and intentional approach. It's time to work cooperatively and collectively through trust relationships rather than working at odds with one another. It's time to share resources and ideas rather than compete for the top prize. It's time to care for Mother Earth as a community, together with innovation through collective modes of thinking.

The Western world has lived a high-consumption lifestyle and attempted to claim environmental achievements in the past few decades. Americans, in particular, have prided themselves in their green steps. Many environmental organizations, trying to motivate Americans, have claimed that a lifestyle change is unnecessary and that we must take at least one green action annually. For instance, start a recycling collection in your home this year, then next year begin composting, then begin a litter collection campaign in the neighborhood the year following. These activities are great! However, our actions need to be more aggressive. We complain about corporations greenwashing their claims of helping the environment, yet most Americans are "greenwashing" their own efforts to mitigate climate change. We need to do more—much more!

We need to rethink the larger picture here. These actions in our individual lives need to add up to change larger systems—we need to make systemic changes—and that starts with each of us. The thinking that "the bigger guys are in control" and should be doing the heavy lifting does not cut the mustard—it never has. All those individual actions I mentioned previously are great activities to start with, but this situation requires significantly more commitment and bigger lifestyle changes. Whoever told you that your lifestyle does not need to change was wrong—*high-consumption lifestyles are exactly what is driving the systemic problem.*

WHAT IF . . . WE RETHINK PLASTIC CONSUMPTION

"Infinite growth of material consumption in a finite world is an impossibility."—**E. F. SCHUMACHER**

Our economy is fueled by consumerism. Note that the measures of economic success we hear daily are all tied to consumer spending. The Dow Jones Industrial Average measures the growth of large U.S. companies through a price index, a metric of consumer and business purchasing. The Gross Domestic Product (GDP) is the estimated market value of all the finished goods and services produced within a country's borders within a year, measuring production related to the demands of consumerism. GDP would fall if consumers demanded fewer goods; likewise, "the sky would fall" (as Chicken Little predicted) to investors. The bottom line is that the national and world economies tank when consumers stop spending.

Return on Investment (ROI)—one of my favorite economic consumption terms—is a business term indicating risk based on time, where investors restrict investment to only projects that provide a quick turnaround of cash, generally within five to seven years. Unfortunately, many social programs investing in the turnaround of human lives do not pass the ROI test, nor do climate and renewable energy projects. Yet I find in my research that plastic production facilities, sometimes called petrochemical facilities, are not evaluated on the traditional ROI test, but rather receive funding through political connections. In other words, these petrochemical facilities require governmental support and cannot financially support their initial construction without state funding.

An example of these failed economics is a petrochemical facility that was to be constructed in Belmont County, Ohio. This plastics manufacturing plant received $100 million in state economic funding. The plant was proposed in 2015; however, after missing numerous construction timelines and delays, the entire project was canceled in 2021.[14] Another similar plant, owned by Shell, to be constructed in northeast Ohio, was scrapped after an initial investment of $300 million, and the state required a refund of its $20 million investment. If standard ROI scrutiny had applied, these facilities would not have been financed.

These plastics production facilities are the tip of the iceberg and the pipeline for the promotion of plastics consumption. As noted earlier, part of the rethink equation is the evaluation of consumers' overconsumption of plastics.

Adam Smith once stated that "consumption is the sole end and purpose of all production; and the interest of the producer ought to be attended to, only so far as it may be necessary for promoting that of the consumer."[15] Yet, over-consumerism is directly related to climate change, and part of addressing the climate crisis is the immediate need to address this consumerism problem head-on. For example, consider the observation made by Mel Gibson that he doesn't believe we are "crumbling as a civilization, but this is not our finest hour, and it's good to be mindful that we're all susceptible to fall and to look at what are the earmarks of a civilization on the wane. What are they—the destruction of the environment? Conspicuous consumption? Heard of those?"[16]

Conspicuous Consumption

"We act as though resources we consume are infinite, and the wastes are invisible."[17] This thought from Brian McLaren brings forward the issue of conspicuous consumption. For example, the global consumption of plastics has quadrupled over the past thirty years, driven by the evolving marketplace. As a result, global plastics production doubled from 2000 to 2019, reaching 400 million tons.[18]

"Nearly two-thirds of plastic waste comes from plastics with lifetimes of under five years, with 40% coming from packaging, 12% from consumer goods, and 11% from clothing and textiles."[19] Approximately half of all plastic produced is "designed for single-use purposes—used just once and then thrown away."[20] Such a significant amount of plastics having a very limited lifespan indicates that consumers are utilizing disposable plastics, increasing the need for more plastics production. Such high consumption rates are not sustainable when you connect plastics to the impacts on the climate, human health, and the local environment.

The term *conspicuous consumption* was first coined by the economist Thorstein Veblen in his 1889 book, *The Theory of the Leisure Class*, which described conspicuous patterns of displaying wealth. Yet, today, we apply the term *conspicuous consumption* to plastics with a slightly different meaning, not implying the display of wealth but rather the frivolity of *convenient disposables*. This is because the current age of plastics has marketed the product as single-use for the user's convenience, hence *convenient disposables*.

The antidote here would be *reduced consumption*, an underrated and missing component of the recycling plastics discussion. That discussion will take place in Chapter 9. But first, how did our society reach our current plateau of conspicuous plastic consumption? How did we get duped into the ongoing marketing strategy of convenient disposable plastics? I believe the answer lies in our primary Pillars of Sustainability.

SUSTAINABILITY REDEFINED

Sustainability has been primarily defined through the Sustainability Pillars, which often drive environmental decision-making for corporations and policymakers. The standard Sustainability Pillars display three concepts to be equally supported: economic, environmental (ecological), and equity (the 3 Es)—sometimes relabeled as Profits, Planet, and People (the 3 Ps). As most businesses and governmental environmental programs are well positioned on the Sustainability Pillars, so are the modern marketing models that produce and market our consumer goods and services. But what if the Sustainability Pillars are misleading us toward a false level of "sustainability"? What if we are not meeting our sustainability goals because we are too reliant on a false model of the Sustainability Pillars? Has anyone professionally critiqued the Sustainability Pillars? Have the Sustainability Pillars been scientifically tested and proven perfectly faultless? I believe not. And who determined that all three pillars must be of equal proportions and importance? I would argue that, looking at where we stand today, the Sustainability Pillars are falsely leading us to depend too much on profits and economic justification in business decision-making. There is not as much emphasis on the environment and equity side of the equation.

Professor Kathleen Smythe challenges the traditional Sustainability Pillars in her book *Whole Earth Living: Reconnecting Earth, History, Body, and Mind.* She notes that the "model was deeply flawed because the dominant economic system was not flexible enough to accommodate the holistic thinking."[21] One of the concepts that Smythe brings out through her discussion of the sustainability triangle is the overreliance on economic development to address poverty. At the same time, the constant pressures of economic growth policies have exploited environmental resources and created a disregard for human welfare. She notes: "A more successful sustainability model will start with who humans are and what they need in order to thrive, to have meaningful work and meaningful lives. With a more realistic view of the limited capacity of technology and economics to meet human needs, the work of crafting a new model can begin."[22]

To pick up where Smythe questions the current Sustainability Pillars, I apply the contrarian viewpoint and challenge this current sustainability model. In addition to concerns about economic dominance driving the model, there is also strong corporate use of sustainability as a marketing tool to sell products and services. The sustainability movement has been "greenwashed" by creative corporate marketers who have commandeered the three-pillar sustainability model to fit their marketing needs. Many corporations offer sustainability annual reports and product sustainability statements, yet very few meaningful changes are offered to greening the product or the package to meet sustainability standards. For instance, most major beverage drink companies have recently been challenged on their broad environmental claims and had to withdraw their public claims.[23]

In addition, when the Sustainability Pillars model is applied, there is often an overemphasis on economics driving the decisions. Economist Herman Daly has said, "There is something fundamentally wrong with treating the Earth as if it were a business in liquidation."[24] Granted, the economics of product development is an extremely important consideration. My point here is that the Sustainability Pillars were not intended to be governed solely by economic decisions. Economics could be one of the evaluation points, thus still a consideration, but not the major driver.

Taking it back as an environmental model concerning the human condition requires reworking the basic tenets of the model, questioning the overemphasis on economics and the underemphasis on human values. The model also needs to reflect where we are regarding climate change and the human impacts on Earth's climate.

To begin with, I suggest that the three Sustainability Pillars be positioned as *four Sustainability Covenants*. Anyone who has sat on a three-legged stool will know its instability. Stability is gained by adding a fourth leg. Therefore, I suggest the exploratory use of the four Sustainability Covenants:

- *Ecojustice Covenant*: The principles of justice, diversity, inclusion, and equity that lift all humans into a respectful societal position.
  - *Action*: We will work to care for life in all its diverse forms, with special care for the vulnerable.
  - *Plastics*: Reduction and elimination of use of plastics offers ecojustice to those living near plastic production facilities, and ensures the safety and health of production facility workers.
  - Sustainable Development Goals (SDGs):[25] SDG 5 (Gender Equality), SDG 10 (Reduced Inequalities), SDG 16 (Peace, Justice, and Strong Institutions), SDG 17 (Partnerships for Goals).

- *Earth Protection Covenant*: A living covenant of Do No Harm to Earth, including protections from greenhouse gases to prevent climate change and actions to increase biodiversity.
  - *Action*: We will work to prevent ecological degradation and encourage ecological restoration activities.
  - *Plastics*: Reduction and elimination of plastics offers reduction of harm to Earth, our common home.
  - Sustainable Development Goals (SDGs): SDG 7 (Clean Energy), SDG 13 (Climate Action), SDG 14 (Life under Water), SDG 15 (Life on Land), SDG 17 (Partnerships for Goals).

- *Education Opportunity Covenant*: The principles of human welfare, individual choices, opportunity equity, and empowerment.

- *Action*: We will work for more equitable access to universal educational opportunities through community collaboration efforts.

- *Plastics:* Reduction and elimination of plastics offers reduction of harm to humans, offering equity and empowerment.

- Sustainable Development Goals (SDGs): SDG 1 (No Poverty), SDG 2 (Zero Hunger), SDG 3 (Good Health and Well Being), SDG 4 (Quality Education), SDG 8 (Decent Work and Economic Growth), SDG 10 (Reduced Inequities), SDG 15 (Life on Land), SDG 16 (Peace, Justice, and Strong Institutions), SDG 17 (Partnerships for Goals).

- *Ecological Economics Covenant*: The concern that every person has food, shelter, clean air, potable water, and employment through sustainable production and consumption, ethical investments, and protection of Earth, our common home.

  - *Action*: We will support a green economy with more equitable access to the economic resources necessary for human life in an eco-ethical manner.

  - *Plastics*: Reduction and elimination of plastics offers food, air, and water free of microplastics, and grants sustainable economic production through green jobs.

  - Sustainable Development Goals (SDGs): SDG 1 (No Poverty), SDG 2 (Zero Hunger), SDG 6 (Clean Water), SDG 7 (Clean Energy), SDG 8 (Decent Work and Economic Growth), SDG 9 (Industry, Innovation, and Infrastructure), SDG 11 (Sustainable Cities and Communities), SDG 13 (Climate Action), SDG 14 (Life under Water), SDG 15 (Life on Land), SDG 17 (Partnerships for Goals).

This is an exploration, and I encourage the experimental use of these Sustainability Covenants. Let the discussion begin.

Artwork by Pam Longobardi + Drifters Project

PLASTIC LOOKS BACK, 2014

Commission for *Sierra* magazine cover, September/October 2014, 28" x 46" x 6". This piece contains hundreds of objects from remote locations in Panama, Alaska, Indonesia, and Greece, including microplastic from Hawaii. The majority of the elements were collected in the extremely remote San Blas region of Panama's Caribbean coast, home to the indigenous Guna Yala peoples. This high impact coastal zone is a primary nesting site for leatherback sea turtles. I worked with the women of the town of Armila for two weeks in a collaboration that involved removing thousands of pieces of plastic from the beach in preparation for the arrival of the leatherbacks and the construction of artwork.

—PAM LONGOBARDI

Chapter 9

LIFE WITHOUT PLASTICS

"You must do the thing you think you cannot do."

—ELEANOR ROOSEVELT[1]

Repair and reuse were everyday activities on Grandfather's farm. Whenever it was suggested that a household item needed to be replaced, it was common for Grandfather to say, "Money does not grow on trees." Then he would take the broken item to his tool barn and work on repairing it. Reusing an object for a second use was expected as well. For example, if a water bucket had a hole in the bottom, it was repurposed into a vegetable-carrying bucket. Old clothes that could not be mended were used as rags or insulation in the barn. Clotheslines that broke were used to tie down lumber piles. The message I received from my grandfather was "waste not, want not." I ran to the library as a teenager to look up that idiom. I found the direct meaning: If we don't waste what we have, we will still have it in the future and will not lack or want it.[2] I wish to add that through reuse, we avoid purchasing new products and thus save on natural resources that might have been utilized.

Revisiting the sustainability covenants from the last chapter, we see that alternatives to convenient disposal, such as reuse, conservation, and reduced consumption—values taught on the family farm—are overlooked in the modern sustainability world. Sustainability is broken when we overlook the obvious alternatives to disposal. But society didn't always think and operate in this mode of convenient disposal, so how did we get here?

During World War II, it was a patriotic duty for citizens to recycle rubber, paper, tin, and other metals for the war effort through scrap drives.[3] Also patriotic was the sense of reuse, conservation, and reduced consumption. The Great Depression (the early 1930s) and the War (most of the 1940s) brought out a spirit of lifestyle best epitomized by the mantra of "use it up, wear it out, make it do, or do without."[4]

Life changed after the war. The message of patriotic duty changed to "consume" as the country transitioned from a war economy to a consumer economy. Consumers in the 1940s into the 1950s assumed that material goods and appliances had long lifespans and could be reasonably repaired. That assumption was challenged with the new plastic-made goods and appliances made after World War II, as plastic-made items inherently could not be repaired.[5] You cannot nail, glue, or screw together broken plastic parts. They need to be replaced.

The conversion to convenience disposables was a rocky road. The "ethos of reuse" was so deeply ingrained in Americans that when vending machines started to dispense coffee in disposable plastic cups in the 1950s, consumers attempted to reuse the cups instead of throwing them away.[6] The public needed some convincing, and so the advertising picked up, with magazine ads promoting the wonders and convenience of disposable plastics. The advertising soon won us over, and thus, we started our addiction to plastics. In the beginning, consumers had to be convinced that disposables were "good." Once people embraced that message, landfills started to fill.

I ask the question: If you do not reuse or repair—if the item is short-lived or disposable after one quick use—do you ever develop an "ownership" of the article? Pride in maintaining it? A sense of responsibility in its disposition? If there is no ownership of the item, then there is no ownership of its disposal, such as by litter, incinerators, or landfills.

In today's world, the fast-paced production of plastic products is designed for one-time use, for lack of ownership care, and for easy, thoughtless disposal. Moreover, I remind you that plastics impact the climate, human health, and the local environment.

BRIDGE BUILDING: A PHASE-OUT APPROACH

The Intergovernmental Panel on Climate Change (IPCC) March 2023 Synthesis Report gives a stark view of reality: Human-caused climate change has resulted in rapid adverse effects across the planet, shifting global weather patterns and climate extremes and leading to widespread damage and losses to nature and people. The report notes, "Limiting human-caused global warming requires *net zero CO_2 emissions*. Cumulative carbon emissions until reaching net-zero CO_2 emissions and greenhouse gas emission reductions this decade largely determine whether warming can be limited to 1.5°C or 2.0°C."[7] Eliminating plastics production and use is a significant part of stopping the warming of Earth's atmosphere.

Choosing a warmer planet is not an option, as continued global warming will result in increased and intensified hazards. "The likelihood of abrupt and irreversible changes increases with higher global warming levels," notes the IPCC.[8]

Binary dualistic thinking would bring us to the thought of reducing our civilization to the pre-industrial age as we eliminate all fossil fuel combustion engines. To have fossil fuels or not. To produce plastics or not. This dualistic thinking was addressed in Chapter 8. To make decisions in a transformational time, it is necessary to move beyond the binary win/loss pathway, stepping into the "contrarian" viewpoint—thinking in a way contrary to the norm. So, again, we need to unfollow the standard thinking patterns and move forward with a new decision pathway.

As I write this book, today is the Fourth of July, America's Independence Day. Many of us celebrate by displaying the Stars and Stripes—the American flag—in front of our homes. Although my flag is made of cotton, most flags in my neighborhood are made of polyester, nylon, or some other form of plastic. It is noticeable from a distance.

Historians note that Betsy Ross hand-stitched the first authorized American flag, made from cotton cloth. Today, of course, flags are machine-made from many materials. Is your flag climate-friendly? Is your flag made from fossil fuels?

I believe that displaying a plastic flag not only fuels our climate crisis but also plasticizes the image of our country!

The same could be said for Great Britain's proud Union Jack, the National Flag of Canada with the red maple leaf, or the National Flag of Mexico with the fierce eagle. Any national flag should be displayed proudly and made with climate-friendly materials.

Where do we move to, as the IPCC warns us of Earth heating up? The obvious next step is to stop using plastics, as we stop producing and utilizing fossil fuels. That hard stop must be soon, as climate scientists' warnings are austere and stark. However, "Plastics are one of the most ubiquitous artificial materials on Earth,"[9] as I noted in Chapter 2. "The global market value of plastics is forecast to grow from 523 billion U.S. dollars in 2018 to more than 750 billion U.S. dollars in 2027."[10]

With such a large-scale global market, plastics will be difficult to eliminate from production quickly. We cannot easily find replacement product materials for the vast uses of plastics. However, fossil fuel production, including using fossil fuels for plastics, will need to be eliminated to address the existential climate crisis.

To ban the production of plastics globally would seem a formidable task. To accomplish this vision, nearly two hundred nations and their citizens would need to work together through an international treaty. Before you write this off as too tall an order, be reminded that we have been down this road before, in different circumstances. There have been successful examples demonstrating that this can be done.

Leaded gas that fueled our vehicles was determined to be poisoning our children, affecting the growth of the human brain. Ingenuity and determination allowed engineers to formulate unleaded fuel and automakers to fine-tune the carburetors and pistons to avoid the need for leaded gas. Between 1986 and 2021, all nations across the world banned lead-based fuel.[11]

In another example, chlorofluorocarbons (CFCs) started being used in refrigeration and air conditioning in 1928 but had the unintended effect of creating an "ozone hole" in Earth's atmosphere. In 1987, an international ban on the production and distribution of chlorofluorocarbons took effect through a 197-nation agreement that was strengthened three years later through the Montreal Protocol.[12]

Hydrofluorocarbon (HFC) refrigerants were primarily used in refrigerators and air conditioners as a bridge for CFC replacement. In recent years, HFC has been considered a greenhouse gas, contributing to the warming of Earth. In 2016, representatives of nearly two hundred countries met in Rwanda and agreed to limit the use of HFC gases in refrigeration.[13] The first bridge for refrigerants was a temporary solution. That happens. A new bridge is required, and global action is underway to build this solution.

These two examples prove that global nations can act in unison when needed. Or as Hannah Richie observes, "Action can happen quickly. Especially when there are large threats to human health on the line."[14] A bridge, or series of bridges, is necessary to move us toward replacing plastics.

Past practices demonstrate that a bridge strategy allows for a transition period for nations, industry, and citizens to adjust and adapt. As a bridge for lead-based vehicle fuel, unleaded gasoline has been utilized worldwide, with some experimentation in ethanol blends and biodiesel. These are fossil fuel-based blends that will need a "clean-energy" replacement. As a bridge to offer time, the U.S. Department of Energy has offered research and development grants in the "renewable gasolines" field as an alternative fuel source.[15] These innovative fuels are developed from renewable sources rather than fossil fuels. They may provide a transitional bridge to supplement the electric (non-combustion) vehicles beginning to be offered on the market. Again, bridges are built through experimentation and innovation.

A bridge for phasing out plastics will take different methods, different actors, and time. The push will come from concerned citizens and leaders calling for the phase-out of fossil fuel production, contrary to the desires of the oil and gas industry. These companies intend to keep business going: There are ramp-up operational plans for new oil rigs for exploration in the Willow Project in Alaska, with planned production of up to 600 million barrels of oil.[16] There are also ramped-up operational plans for new plastic production plants, such as the $6 billion ethane cracking plant Shell is building in Monaca, Pennsylvania.[17]

The real push for a bridge away from plastics will come from educated citizens and national leaders concerned about Earth's warming. We are in an existential crisis. I propose a six-step bridge toward the goal to eliminate the production and use of plastics. This bridge is focused on the leadership and advancement of the industrialized nations, and on the continued efforts of the Intergovernmental Negotiating Committee toward adoption of the Global Plastics Treaty.[18]

STEP 1: BAN NEW VIRGIN PLASTIC PRODUCTION IMMEDIATELY

"If climate change is our planet's disease, global warming is one of the pathogens causing it."—**ERIN LINKO**[19]

In utilizing this medical metaphor, the first step of bridging to a life without plastics is to stop the bleeding. The production of plastics must stop now. Authors from the Center for International Environmental Law predict that if we remain on our current projection of plastic production, "by 2030, these emissions [from plastics production and transport] could reach 1.34 gigatons per year—equivalent to the emissions released by more than 295 new 500-megawatt coal-fired power plants. . . . By 2050, the cumulation of these greenhouse gas emissions from plastic could reach over 56 gigatons—10–13 percent of the remaining carbon budget."[20]

The IPCC's Sixth Assessment Report calls for reducing greenhouse gas emissions to net zero by 2050.[21] This cannot be achieved with continued plastic production.

I propose numerous steps to help humans reach net zero, but the first must be the universal worldwide ban on producing new virgin plastics. Stop the harm to future generations. Stop producing new plastics.

The CIEL reports that "production capacity is expected to increase by 33–36% for both ethylene and propylene" by 2025. In addition, the CIEL report notes that "this massive expansion in capacity could lock in plastic production for decades, undermining efforts to reduce consumption and reverse the plastics crisis."[22] Regarding planned or new plants, the report states that there are "massive new investments in plastics infrastructure in the U.S. and abroad, with \$164 billion planned for 264 new facilities or expansion projects in the U.S. alone."[23]

From a climate perspective, there is an immediate need to cancel all new start-up production of plastics. The shutdown of the new proposed plants is based on the extreme climate impacts and human harm. As documented in Chapter 4, plastic and production facilities have hazardous effects on human health and the environment. Chapters 1, 4, 5, and 6 document plastic's impacts on Earth's climate.

The Basel Convention may be an excellent parallel situation on which to model this phase-out of plastic production. The goal of the Basel Convention is "to protect human health and the environment against the adverse effects of hazardous wastes." Furthermore, the principal objectives are:

- "The reduction of hazardous waste generation and the promotion of environmentally sound management of hazardous wastes, wherever the place of disposal;

- the restriction of transboundary movements of hazardous wastes except where it is perceived to be by the principles of environmentally sound management; and

- a regulatory system applying to cases where transboundary movements are permissible."[24]

The national state party signatories to date on the Basel Convention total 190,[25] indicating that the vast majority of the world's nations have embraced the principles and requirements of the Basel Convention. This same model of world agreement is needed to eliminate the expansion and growth of the plastics industry for the protection of human health and the environment against the adverse effects of plastics production.

The effort to stop producing new plastics through newly constructed plants can be achieved as an attachment to the Basel Convention or a freshly created worldwide agreement through the United Nations' unified efforts to combat climate change. As the most recent IPCC report calls for a unified goal of *net zero CO_2 emissions*,[26] and plastics production does involve massive amounts of CO_2 emissions, there needs to be a call for the elimination of these proposed new plastics plants to address the ever-increasing warming of our home, Planet Earth.

>⬦⬦⬦⬧<

Here is a list of experts that citizens can engage with to support their efforts to *ban virgin plastics production*:

Halt the Harm Network—halttheharm.net

People Over Plastics | Climatebase— climatebase.org/company/1134410/people-over-plastic

Plastic Free Future—plastic-free-future.org

Third Act—thirdact.org/our-work

Beyond Plastics—beyondplastics.org/act

STEP 2: REDUCE THE CONSUMPTION OF PLASTICS

As noted in Chapter 6, the overproduction and overconsumption of plastics is an unsustainable lifestyle impacting Earth's carrying capacity, as measured through fossil fuel consumption and the rise of Earth's global temperature. Plastics were declared "toxic" under Canada's primary environmental law, as I discussed in Chapter 7, leading to the rethinking of our consumption habits in Chapter 8. This now leads to the discussion of *reduced consumption*, an underrated and missing component of the current discussion around recycling plastics.

Our sustainability discussions at present exclude reducing consumption, primarily because of the broken definition of sustainability and its heavy reliance on new growth economics. (See Chapter 8 for my proposal on redefining the Pillars of Sustainability.) Reducing consumption challenges the current economic equations, yet sufficient evidence shows that plastic replacements will offer entrepreneurs new economic opportunities. Recently, the United Nations Conference on Trade and Development estimated that the global trade value of plastic substitutes stood at $388 billion.[27] The point I desire to raise is the overreliance our society has placed on plastics, particularly disposable plastics.

Alongside the consumption rate of plastics, global plastic waste generation doubled from 2000 to 2019, reaching 353 million tons.[28] The sheer volume of single-use plastics and the lifestyle of conspicuous consumption leads to the discussion of the need for "reduced consumption," an underrated and missing component of the international recycling plastics discussion. Consider the children's book *Charlie and the Chocolate Factory* by Roald Dahl. An exciting interpretation of this book by Malu Rocha notes that key characters in the book emulate "the standard economic model of consumption," which "invokes the sort of desires that motivate "the characters."[29] Four kids are presented as greedy, displaying various forms of conspicuous consumption. Through various obstacles, they do not win the grand prize. Charlie, presented as humble, polite, and unselfish, eventually wins the unknown prize—a moral lesson on the ugly greedy side of consumerism.

Reduced plastic consumption begins with breaking down the barriers of assumptions. Do we need to wrap to-go eating utensils in plastic? Is it more sterile and hygienic? The answer is NO! The hidden part of the plastic wrap industry is that its facilities are NOT sterile and do not present themselves as "safer" environments for packaging food-ready items or eating utensils. It is just an assumption presented with persuasive advertising and amplified during the fear of the pandemic.

Upstream, a nonprofit working toward solving the plastics problem by shifting people, businesses, and communities toward reuse, has unmasked the plastics industry myths, demonstrating that they do not produce sterile products.[30] I have toured several plastic packaging plants and have witnessed the factory conditions. Single-use plastic product manufacturing facilities are not white-glove "clean manufacturing" processes. They are machine and human hands-on factories that do not guarantee sterile or antiseptic conditions. I encourage you to do a factory tour to investigate. Bust the myth and change your attitudes toward single-use plastic wraps.

A positive example of governmental action to reduce consumption is the Canadian effort to reduce single-use plastics through legislative policy. One week before the 2022 UN conference, Canada's federal government announced a new policy to ban "companies from importing or making plastic bags and takeout containers by the end of this year, from selling them by the end of next year and exporting them by the end of 2025. The move will also affect single-use plastic straws, stir sticks, cutlery, and six-pack rings used to hold cans and bottles together." Environment Minister Steven Guilbeault stated that the Canadian government is "targeting 2030 to eliminate all plastic waste from ending up in landfills or as litter on beaches, in rivers, wetlands, and forests."[31]

Reducing plastic consumption can and should happen at all scales, in the workplace, schools, governmental settings, and at home. I often see the plastics industry pointing the finger at the consumer: "We are just making plastic because that is what people want." Starting at the consumer level, what you do sends a loud message to the plastic producers.

Simply replacing disposable plastics with durable, reusable items will go a long way toward reducing the waste stream, the litter flows, and the overreliance on plastics. Moving further along that trail, take a long look at your use of plastics on an average day. Notice the amount of plastic packaging waste in the grocery, the plastic wrap around plastic ware at the carryout lunch counter, the plastic bag the cashier stuffed your two small items into at the pharmacy, and the plastic pen you hold to write a note on a greeting card that was plastic wrapped when purchased.

Take an inventory of the plastics in your life. List the single-use plastics and the durable plastics you touch in a day. Start first on the single-use plastics and systematically replace them, one by one, with items that are not made of plastic, are more durable and reusable, and can last through many uses. An easy start is a grocery or carry bag. Then, move on to office supplies. A significant step would be to tackle the kitchen and eliminate plastics from food contact.

Where do you start toward reducing plastics beyond your home? It might seem that you are a small cog in the large world of consumerism. However, we all have the purchasing power of organized consumers, as realized by César Chávez during the mid-'60s when he organized a nationwide grape boycott to benefit workers who picked grapes in deplorable working conditions and were grossly underpaid. The success of the nationwide consumer boycott placed tremendous pressure on grape growers, who eventually signed a labor agreement with the National Farm Workers Association that offered better working conditions and fairer pay structures.[32] Consumers have since launched many successful boycotts. Why not organize a consumer boycott effort to reduce plastic production?

Where would a campaign to reduce plastic consumption start? How about the very businesses that promote single-use plastic consumption?

We can push beverage companies, fast-food chains, and large brand owners of plastic products to look inward at their processes, packaging, and practices to reduce waste. The World Wildlife Fund released a report on various companies and "believes that businesses are uniquely positioned to reduce waste within their supply chains through better sourcing, improved design, and business model innovation." The WWF study estimates that "100 companies have the potential to prevent roughly 50 million metric tons of the world's plastic waste by 2030."[33] The report features the top nine plastic waste companies, which include the Coca-Cola Company, Amcor, Proctor & Gamble, Colgate-Palmolive, Keurig Dr Pepper, McDonald's, Starbucks, Kimberly-Clark, and CVS Health. If you happen to work at one of these companies, you could propel a better supply chain design. If looking from the outside in, how about petitioning these big-impact companies to make positive changes? Let us start a consumer awareness drive to reduce the consumption of plastics.

>◇◇◇◇◇<

Here is a list of experts that citizens can engage with to support their efforts toward *waste reduction of plastics*:

Beyond Plastics—beyondplastics.org/act

Plastic Ocean Project—plasticoceanproject.org

Oceanic Society—oceanicsociety.org/global-ocean-cleanup

Zero Waste International Alliance—zerowaste.org

Surfrider Foundation Mission—surfrider.org

STEP 3: REUSE OF RECOVERED PLASTICS

Moving along the bridge, the third step is the reuse of plastics. This is an often-overlooked concept as people mistakenly blend reuse with recycling. *Reuse*, by definition, "means to extend the life of a product, package or resource by either using it more than once with little to no processing (same or new function), repairing it so it can be used longer, and sharing, renting, selling or donating it to/with another party."[34]

The "throw-away" economy has significantly contributed to ever-increasing carbon emissions. The "CO_2 emissions from disposable paper, plastic, and polystyrene cups are 3 to 10 times greater than those of reusable ceramic, stainless steel, and glass when compared over their life cycles."[35] Reuse can directly reduce greenhouse gas emissions by trading out single-use or short-term use plastics for nonplastic alternatives.

Reuse can take many forms, but it is distinctively different from recycling. Reuse can be in the form of using a grocery bag many times over, replacing single-use plastic bags. Cloth bags can be washed with your towels, thus cleaning the bags without the "extra" environmental burdens of caring for them (the towels need to be washed anyway; add the bags to the load). I highly discourage using thicker "reusable" plastic bags as they add to the fossil fuel climate load. As you unload your car of bagged groceries, do not forget to reload your vehicle with empty reusable bags for the next shopping trip.

Reuse can also take the form of refillables, such as refilling a durable and reusable water canister to replace the "need" for single-use plastic water bottles. In 2018, Americans bought more than 70 billion plastic water bottles, with three out of four ending up in a landfill, incinerator, or illegally littered. Why? The answer is primarily convenience. Moving from disposables to refillable containers takes the smallest amount of planning ahead, but will address a significant part of the plastic disposal problem and reduce the fossil fuel climate load.

The U.S. Plastics Pact's Target 2 states: "100% of plastic packaging will be reusable, recyclable, or compostable by 2025."[36] In pursuit of that goal, the Pact held a webinar in May 2023, displaying several innovative refillable programs introduced in the United States and Europe. The Pact's annual awards program celebrates the "Reuse & Refill Sustainable Packaging Innovation Award Finalists."[37] These innovations could be next year's standard retail offerings of refillable programs for drinks, soaps, food products, and more.

Closed Loop Partners' Center for the Circular Economy, in partnership with the U.S. Plastics Pact, released a reuse report for the retail industry in March 2025. The report, entitled *Getting Ready for Reuse in Retail*, offers five product categories that are ready for the near-term implementation of reusable packaging: prepared food packaging in retail, fresh produce containers, beverage bottles for localized supply chains, home care product bottles, and personal care product bottles.[38] The report authors offer insights to established retailers and entrepreneurs for developing sustainable reuse businesses.

The U.S. National Park Service prohibits single-use plastics such as straws, bottles, and bags from vendor contracts. "The directive [from Director Deb Haaland], which follows an executive order on sustainability that President Joe Biden signed December 2021, would phase out the sale and use of these items in national parks, wildlife refuges, and other public lands by 2032."[39] The long phase-in period is due to U.S. law forbidding the overlay of new regulations on existing contracts, requiring the new rules to be phased in at the renewal or rebidding of service contracts. Unfortunately, this effort was reversed by a 2025 presidential executive order.

There are additional benefits to eliminating single-use plastics, as noted by the National Parks Conservation Association. "Reducing our dependence on these petroleum-based products not only benefits the climate, it lightens the burden on park waste management systems, decreases litter on lands and waterways, and helps protect wildlife from plastic ingestion," said Sarah Barmeyer, NPCA's senior managing director of Conservation Programs.[40]

My favorite reuse award is The Reusies®, sponsored by the nonprofit Upstream, which recognizes annually the Most Innovative Reuse Company of the Year, Corporate Initiative of the Year, Community Action of the Year, and Activist of the Year. As Upstream's website puts it, "The Reusies® celebrates the pioneers, the trailblazers, the innovators, and game-changing heroes who are developing a better way than a throw-away, advancing systemic change and co-creating a world where we can get what we need and want without all the waste."[41]

Reuse is also represented through the repair of items. Do-it-yourself repairs might seem daunting, but repair clinics offer approachable instruction and a way to rely more on durable items than disposable plastics. There is a national trend toward hosting fix-it clinics in local libraries, community centers, and schools. Whether you're handy or trying to become more so, repair clinics are easy to organize, as do-it-yourself repairers in every community are ready to support a repair event. Fixit Clinic offers training videos on its website on starting a repair clinic. Their mission is "Education, entertainment, empowerment, elucidation, and, ultimately, enlightenment through guided disassembly of your broken stuff."[42]

Finally, let's consider the restaurant trips, where you take home your leftovers from the too-large-to-eat-in-one-sitting dinner. We often come home with restaurant-supplied containers made from—plastic! Not just plastics, but usually nonrecyclable, nonreusable plastic! A couple of friends of mine have taught me a good lesson to avoid this distress. They come to the restaurant with a reusable carrying bag containing several clean reusable containers from their kitchen for the dinner leftovers. I admire their effort and now try to model it. Imagine all the plastic avoidance through the reuse of your own dinnerware.

Some vendors are catching on to this concept, with easy-to-carry reusable containers you can pack for vacations and travel trips. Many large sporting goods retailers carry them, and they may be found at some eco-shops. The *Treehugger* newsletter lists seven reusable utensil sets in their June 2022 edition.[43] I add that reusables need not be trendy, stylish, or expensive; they need only be functional and nonplastic.

><><><

Here is a list of experts that citizens can engage with to support their efforts toward the *reuse of plastics*:

Upstream—upstreamsolutions.org

Fixit Clinic—facebook.com/FixitClinic

Helpsy: Clothing Reuse—helpsy.com

Rheaply—rheaply.com

iWastenot Systems—iwastenotsystems.com

U.S. Plastics Pact—usplasticspact.org

Refill Madness—refillmadnesssacramento.com

Re:Dish—redish.com

STEP 4: RECYCLE EXISTING FLOWS OF PLASTICS

My entire forty-five-year career was in the promotion of recycling. I still believe in recycling, but I have grown skeptical of plastics. As we discuss the recycling of plastics, the number tossed around is a 9 percent plastic recycling rate, despite heavy investments in public education and attempts to capture plastics for recycling in nearly every community. My friend Stephen Alexander, president/CEO of the Association of Plastic Recyclers, strongly disputes the 9 percent plastics recycling rate. Steve notes that the debate is "misleading" as the authors of the referenced articles "failed to acknowledge that the low numbers they cite include ALL plastic items, including durable plastic items not ever collected through community recycling programs."[44] I agree with him that numerous national and international publications have been misleading and generally seem to work against the proper education of the public to recycle accurately.

The 9 percent figure represents the approximate recycling rate of ALL plastics generated. This book is about all plastics, not just the traditional recyclable plastics. Not all plastics are created equally, and not all are intended to end up at a recycling facility. However you slice the pie and generate a plastics recycling rate, the bottom line is that more needs to be recycled, and most plastics get landfilled or illegally disposed of. Can we solve the problem by turning the heat up on the recycling industry? Not really. As the plastics recyclers themselves will readily admit, there is a slice of the pie they prefer to see recycled, including consumer beverage containers and a limited amount of clean commercial and industrial stream of plastics. Plastic durables such as plastic refrigerator parts, air conditioner shells, computer casings, auto parts, sports equipment, etc., are not included in the "desired" recycling streams. Also omitted are plastic items such as razor casings, toothbrushes, dishes, utensils, nylon rope, eyeglasses, and other small plastic items you might find around the house. All these are missing from your recycler's list.

Raising the recycling flag and recycling more plastics that could be recycled is part of this bridge strategy. But let's not kid ourselves or greenwash our recycling efforts. To be clear, we can't recycle our way out of the plastics crisis. Plastics cannot be indefinitely recycled. A plastic bottle can be recycled into a plastic bottle (and other items), but not many times over. Degradation, contamination, and material loss take their toll on the recycling process, and there will be an eventual end point to the recycling trail. However, recycling does have its environmental benefits over other disposal methods.

Another concern with recycling plastics is the exploitation of shipping unwanted plastics (and wastes) to underdeveloped nations, utilizing severely low-paid contractors to sort and clean plastics of various grades. The issue was highlighted in the trade disputes between China and the United States, as I described in Chapter 6; however, there were significant spillovers to dozens of additional countries.

Beginning in 1989, countries worldwide gathered to work on an international agreement regarding the transboundary movements and disposal of hazardous wastes. In 2019, these nations added plastics as a "regulated waste" to the Basel Convention. As of 2023, 191 nations have signed this treaty; however, the United States is not a party to the treaty. Despite that, export shipments of plastic waste from the United States are now considered "criminal traffic as soon as the ships get on the high seas," according to the Basel Convention.[45]

Thus, shipping plastics across the seas for recycling is now prohibited, unless certain provisions are granted, requiring the rebuilding of U.S. domestic recycling infrastructure. This has been a challenge and a need within the North American recycling system, as the older recycling systems need financial restructuring to better capture the various plastics and other commodities coming in through residential and commercial recycling streams.

One domestic recycling infrastructure that has proven to work in ten states is state bottle deposit legislation, often referred to as bottle bills. Recovery rates of beverage containers typically are around 71 to 90 percent depending on the deposit required and the consumer convenience factors.[46] Plastic bottles, as well as other beverage containers, have the highest recycling rates in these states, presumably due to the container deposits that drive consumers to return the containers. The Can Manufacturers Institute, Glass Packaging Institute, and Association of Plastic Recyclers have lobbied for new beverage container refund programs and support existing bottle collection infrastructure. There is also support for a national beverage container refund program.

How can the recycling industry support transitioning away from the fossil fuel climate load? To start with, the U.S. Plastics Pact offers four target goals:

- Target 1: Define a list of packaging that is problematic or unnecessary by 2021 and take measures to eliminate items on the list by 2025.

- Target 2: 100 percent of plastic packaging will be reusable, recyclable, or compostable by 2025.

- Target 3: Achieve an average of 30 percent recycled content or responsibly sourced biobased content in plastic packaging by 2025.

- Target 4: Undertake ambitious actions to effectively recycle or compost 50 percent of plastic packaging by 2025.[47]

To achieve these ambitious goals, the U.S. Plastics Pact has engaged workgroups to create innovative solutions. The Pact comprises more than 120 "Activators" representing trade associations, retail businesses, recycling converters, recycling reclaimers, raw material suppliers, material recycling processors, consultants, technology innovators, and not-for-profits.

Note the effort to encourage reuse, recycling, and composting through Target 2, working with the plastic packaging industry to fit their product into one of those three end-of-life activities. This is a game changer in packaging design, to design out disposal. The Target 3 workgroup aims to reintroduce recycled content into plastic packaging to extend the life of existing plastic. However, the efforts of the Target 4 workgroup to recycle or compost 50 percent of plastic packaging will require a significant upfront redesigning of the packaging and hopefully achieve a reduction in the use of plastics.

Ocean plastics are one form of plastic that can be captured and recycled yet receive very little attention. According to Ocean Titans, 6.5 million tons of plastics are disposed of in the oceans globally each year.[48] Ocean plastics are plastics disposed of or abandoned in the marine environment; most originate as litter or dumping from land sources. Various groups are working on retrieving and recycling these plastics, including the Oceans Plastics Recovery Project,[49] a three-man team focused on researching and exploring the depth of ocean plastics, the collection complications, the reclamation issues, and potential recycling opportunities. Oceanworks is another company recovering HDPE PCR (Post Consumer Recycled) material captured and collected from floating ocean "gyres." They then market this material to their client base of plastic suppliers—a supplier network "with a capacity of over 400 million lbs. annually."[50]

However, there are significant challenges to recovering ocean plastics. As plastics enter the ocean, many grades of plastics are mixed, degraded, pulverized, chopped, and shredded. The smaller the pieces, the more challenging they are to capture. In addition, the smaller the pieces, the more dangerous the plastics are to marine life through ingestion. Seawater is also extremely salty and has a pH value of around 8.0, leaning it to the alkaline side, making it chemically abrasive to some plastics. This ocean environment makes it difficult to recover ocean plastics and unlikely that those recovered will be fit for recycling.

Raincoast Conservation Foundation senior scientist Peter Ross says that plastic recycling will not reduce the waste stream into oceans and that recycling plastics as an answer to ocean plastics is fiction. "The reality is recycling is not going to be the panacea that saves the world's ocean," he said.[51] The answer lies in preventing pollution of our waterways and oceans. We cannot recycle our way out of this plastics problem or its impacts on the climate and Planet Earth.

⋊⋉⋊⋉

Here is a list of experts that citizens can engage with to support their efforts to *recycle plastics*:

National Recycling Coalition—nrcrecycles.org

Oceans Plastics Leadership Network—opln.org

Sustainable Ocean Alliance—soalliance.org

Global Plastic Action Partnership—globalplasticaction.org

World Wildlife Fund—worldwildlife.org

Earth 911—earth911.com

STEP 5: INTRODUCE PLASTIC SUBSTITUTES

"SKEUOMORPH

Part of speech: noun

Origin: Greek, late 19th century

An object or feature which imitates the design of
a similar artifact made from another material."

—OXFORD ENGLISH DICTIONARY[52]

In 2010, the American Chemistry Council estimated that about 8 percent of global oil and gas supplies were used as feedstock for the production of plastics[53] A 2018 report from the International Energy Agency estimates that 12 percent of the total demand for oil in 2017 was for the production of plastics, accounting for 12 million barrels per day.[54] The report predicted a 100 percent increase (doubling) over the next twenty years due to the increased use of plastics. Using oil (and gas) to produce plastics proportionally increases CO_2 emissions, directly extending the existential climate crisis.

It is time to invest our energies into plastic substitutes—skeuomorphs—to reverse this trend of climate impacts. There is sufficient evidence to note that replacements for plastics will offer economic opportunities for entrepreneurs. Consider the business opportunities for expansion into replacement products utilizing substitute materials for plastics. The caution is not to replace plastics with something equally harmful to the environment. *Regrettable substitution* is when a substitute material with an unknown or unforeseen hazard is used to replace a known hazardous material identified as problematic and needing replacement. Again, the caution is to seek safe substitutes.

I previously mentioned the efforts of the U.S. Plastics Pact. Consider their ambitious Target 1 to define a list of problematic or unnecessary packaging. That list has been defined and was voted on in 2024 to take measures to eliminate these items from production and distribution within the U.S.:[55]

- Plastic cutlery
- Plastic coffee stirrers
- Plastic drink straws
- Non-compostable plastic packaging
- Foamed PET
- Intentionally added PFAS
- Multi-material plastic packaging
- Multi-material film or bags
- Multi-material film or bags for individually wrapped items
- Multi-material pallet stretch wrap
- Multi-material bread bags
- Multi-material cereal bags
- Multi-material bags for fresh and frozen fruit
- Multi-material bags for fresh and frozen vegetables
- Multi-material rigid plastic packaging
- Non-compostable produce stickers
- Carbon black plastics
- Opaque or pigmented PET
- Oxo-degradable additives to plastics
- PETG—a glycol additive in rigid packaging
- Problematic labels (e.g., adhesives, inks, materials)
- Polystyrene (PS, EPS)
- Polyvinyl Chloride (PVC, PVDC)

There is now a void to be filled in finding plastic substitutes for some of these categories—again, noting to avoid the problematic issues of single-use and nonrecyclable or nonreusable materials. Many items on this list are problematic for the reuse and recycling industries and should be phased out permanently. Yet some of these products could be replaced with new materials, which will take creativity and entrepreneurial spirit to develop.

The potential for plant-based cellulose to make a comeback is a game-changer. Before the invention of modern-day chemical fossil fuel plastics, cellulose was the standard material utilized for what is now labeled plastic—cellulose combs, hairbrushes, pens, electrical wire coatings, toothbrushes, razor handles, and clocks. Cellulose was a commonly used material in the early twentieth century. With modern electronic machinery, manufacturing new products using cellulose can be explored.

Loosely defined, cellulose is an organic material found naturally on Earth in the forms of cotton, wood, and hemp. It is a natural, renewable material that offers little to no supply chain difficulties, as it can be grown as a cash crop, although sustainability concerns are raised in the manner and resources utilized.

The Grenoble INP Foundation created Cellulose Valley to research and develop new and innovative cellulose uses to replace fossil fuel plastics. Their primary focus is the "development of alternative solutions, in particular, to single-use food and cosmetics packaging."[56] Some of the challenges these researchers have taken on include "cheese and chocolate packaging, cushioning materials, new types of corrugated cardboard, and high-performing molded cellulose. The objective is to work on primary and secondary packaging . . . that share the same problems, like the barrier function for liquid formulas, closing constraints, or decorations."[57]

While cellulose packaging is still in the development phase, Cellulose Valley is already consulting closely with industry players to produce alternatives that could really make it to market as we phase out plastics. Their efforts are "supported by a large number of industrial partners, enabling it to cover the essential points of the value chain associated with cellulose products and to propose solutions in line with the real challenges facing the packaging sector."[58] Their innovations not only show promise for a future without plastics, but have active support from companies and businesses to walk along this new path. For these innovations to have an impact, the support of companies is key.

A toy company synonymous with plastic has committed to replacing its plastic components with alternative substitutes. In 2014, LEGO launched a search for alternatives for its recycled polyethylene terephthalate (PET) plastic building blocks. "So far, Lego has tested more than 600 materials as options, including bio-polyethylene, which it uses to make botanical elements and Minifigure accessories. In 2023, 18 percent of the resin it purchased for its bricks was made from renewable or recycled sources mixed with virgin sources."[59]

Another emerging plastic substitute is *bioplastics*. Unlike traditional plastics made from fossil fuels, bioplastics are Earth-friendly. The Plastics Industry Association defines bioplastics as plastics that are 1) biobased, meaning they come from a renewable resource; 2) biodegradable, meaning they break down naturally; or 3) both biobased and biodegradable. For example, there are durable bioplastics made entirely from sugar cane.[60] Bioplastic can also be made from food waste, vegetable oils, corn starches, or agricultural field waste products. The Plastics Industry Association also notes that "globally, over 1.7 million metric tons [of bioplastics] were produced in 2014 and contributed to $4.4 billion and 32,000 jobs in the U.S."[61]

There are three immediate advantages to utilizing bioplastics. First, it is a renewable product that can be replaced through a natural cycle.

Second, if not contaminated, it can be composted at its end-of-life. Third, it is not the product of fossil fuels and can be produced through renewable energy sources.

There are experimental products commercially available today utilizing bioplastics technology. Unfortunately, these new products have two significant faults. First, they rely on traditional manufacturing production lines for cost efficiencies, meaning they rely on fossil fuel energy inputs. Thus, the transition to renewable energy sources is necessary to fully transition bioplastics as a complete substitute to plastics as an answer to the climate crisis.

The second issue is that most initial bioplastic products are "food-oriented, single-use applications, such as utensils, straws," and bio-plastic bags.[62] Most plastic food packaging is dedicated to single-use applications, sometimes with the added complication of multilayer, metallic-lined bags for the freshness of the product, and bio-based plastics are adding to that lineup. Although the intention is to collect these items for composting, the problem is generating more waste from *single-use applications*. To entertain a new substitute for traditional plastic, the product or packaging needs redesign to avoid the issues that lead to waste and contamination.

The company Mars Wrigley is experimenting with bioplastic for packaging needs, primarily for its candy products. The trial packaging will have "Nodax PHA (polyhydroxyalkanoate), a biodegradable alternative to petrochemical plastic."[63] The company claims that PHA will "biodegrade in a matter of months in an industrial compost facility." Mars Wrigley is working to develop this new bioplastic in collaboration with Danimer Scientific. This PHA was developed "through a fermentation process involving soil and plant oils." As the package is discarded and composted, it returns to carbon dioxide, water, and biomass. Plants absorb the carbon dioxide, and we extract that carbon in the form of vegetable oil."[64] The company plans to scale up to more product lines after successful deployment with one product package.

Another bioplastic making the rounds of experimentation is PLA (polylactic acid), which is derived from the fermentation of sugarcane or starch from corn, cassava, or potato plants. Some PLA plastic products have added chemicals and resins to make the packaging look more like plastic.[65] These added chemicals may cause the packaging to perform better but may make it non-compostable.

Many municipalities and states in the United States are looking for professional guidance on what is (and is not) compostable or recyclable plastic. Many compostable PLA products have been inspected and certified by the Biodegradable Products Institute (BPI), a nonprofit organization that runs independent tests of compostable products to verify that they genuinely biodegrade or break down into safe, natural components.[66] As this is an emerging field, there is still much uncertainty regarding the placement of PLA in the packaging world. Many feel its best placement is in food packaging, leading to discarded packaging being composted rather than recycled.

There is a raging debate about whether PLA beverage cups, distributed at football and baseball stadiums across America, should be collected for recycling or composting. There is no national answer. Most recycling MRFs cannot sort PLAs, nor is there a nationwide recycling market to consume this type of bioplastic.

Likewise, on the composting end, PLAs cannot be composted in your backyard composter, as they take too much time, pressure, and heat to decompose. Commercial composters generally require a sixty- to ninety-day turnover of their compost piles. PLAs do not degrade sufficiently in that short period. Most commercial composting operators reject PLAs as they are not set up for the conditions needed to break down the bioplastic, and they are suspicious of look-alike plastic packaging that is not compostable. Biocycle researched these issues and "found that only 49 out of 4,700 commercial composters nationwide accepted compostable plastic products."[67] That may be a changing storyline over the next decade.

Additional unanswered issues about PLA are at the front end of the production line—the growing of the crops utilized to manufacture PLAs. Fertilizers, pesticides, and water in drought regions are a concern for growing these crops, raising additional sustainability questions. Eunomia Research & Consulting, in collaboration with the Plastic Pollution Coalition (PPC), published a harsh review of PLAs' environmental and social effects.[68] Key findings of the report and supplement include:

- PLA is an industrial bioplastic. It is made from corn or sugarcane grown with intensive agricultural practices, which contribute to ecological issues such as deforestation, water pollution, soil degradation, and loss of biodiversity.

- PLA is produced industrially in large facilities and with transportation networks that pollute air, lands, and waters and drive environmental injustices.

- For PLA and other industrially made bioplastics to decompose and biodegrade, they must be collected and composted in one of the limited number of carefully controlled, high-temperature industrial composting facilities in the United States that accept them. Such composting facilities are few and far between, and in the absence of appropriate recycling infrastructure, PLA products in the United States largely follow the same waste streams as regular plastics, accumulating in landfills and contributing to plastic pollution.

- PLA and other industrial bioplastics have been found to contain as many as twenty thousand chemical features, indicating a broad range of chemical additives, many of which are identical to those added to conventional plastics.

- PLA does not biodegrade at the same speed as other organic materials in composting facilities, which can lead to contamination of the final compost product.

- When PLA is landfilled, it may fragment into chemical-laced microplastics that pervade our environment and contaminate food and water sources.[69]

In short, we cannot let ourselves be fooled by marketing to believe that certain plastics are biodegradable and fine for the environment when, in reality, the harm remains. As another example, certain laundry/dish detergent pods advertise to be "plastic-free," and note that they don't generate microplastics, are not sold in plastic bottles, and have no plastic ingredients. The "fact sheet" states that the entire product will dissolve in water. However, beware of a wolf in sheep's clothing. Check deep into the listing ingredients. What you are likely to find is an ingredient called PVA, or Poly-Vinyl Alcohol. This plastic product is a thin external coating containing the detergent pod (dishwasher use) or the detergent sheet (laundry machine use).

Many studies "suggest that the degradation of PVA occurs under a specific set of circumstances, which may not be ubiquitous within WWTPs [wastewater treatment plants] or the natural environment."[70] Certain studies note that the PVA is biodegradable in the presence of certain bacteria under lab testing. Yet this is not readily replicated in real-world situations. In addition, most wastewater treatment and drinking water facilities cannot filter out PVAs that have not fully degraded. In other words, the product may "dissolve," like salt in water, but it does not disappear through complete biodegradation. PVAs are seeping into the natural waterways, and eventually into our drinking water.[71] The point is that this PVA coating is a plastic product made from fossil fuel production, and plastics do not biodegrade without the assistance of certain lab conditions. Detergent pod and sheet manufacturers utilizing PVA should not greenwash their products with the label "plastic-free."

What happens to this plastic from PVA-coated detergent pods or the detergent sheets after consumer use? The dishwasher utilizes the detergent in the pod but discharges the PVA into the wastewater. Likewise, the washing machine utilizes the detergent in the washing sheet and then discharges the PVA into the wastewater. Now your local wastewater treatment facility (at the cost of local taxpayers) must screen out the PVAs before they hit local rivers and streams. However, wastewater treatment facilities are not designed to filter out this class of chemicals. Researchers at Arizona State University studied this issue and found that "as much as 75% of PVA goes untreated in the U.S. each year. That amounts to about 8,000 tons of the plastic material being released annually onto land and into waterways across the country." The study concluded that "it's time to take a closer look at what it actually means for a material to be biodegradable, and whether companies should be allowed to claim their products are biodegradable if they only are so under specific conditions."[72]

Although bioplastics have a long road to travel before replacing fossil fuel plastics, Project Drawdown lists bioplastics as one of its crucial climate solutions. Project Drawdown's studies "model the growth of bioplastics to capture 89–100 percent of the [plastics] market by 2050, avoiding 1.33–2.48 gigatons of emissions."[73]

As we phase out plastics due to their impact on the climate, we must find replacements that are not harmful to the environment and human health based on adherence to the precautionary principle. That will require more research and development on bioplastics as a substitute for fossil-fuel plastics. I am reminded of a parallel metaphor as we search for plastic substitutes. I own an electric car, and as an "early adopter," there are challenges to long-distance travel in finding charging stations when needed. I may travel 250 miles on a single charge, but will there be a charging station I can use to recharge when needed? Phone apps are beneficial for planning purposes, but in these early days of electric vehicles, some states have dead zones with no public charging stations. Likewise, there is a search for plastic substitutes happening before the sincere effort to decrease fossil-fuel-generated plastics. In both cases, we must count on society, businesses, and local infrastructures to support the needed changes.

I have noticed that most of the newer branded plastic substitutes on the market are single-use products. As consumers, we need to demand durable goods and migrate away from the wastefulness of single-use products. As we phase out fossil fuels and get serious about phasing out plastics, opportunities will arise for new product lines. Plastic substitutes may have hurdles to overcome, yet great opportunities await entrepreneurs with vision and innovation.

><><><

Here is a list of experts that citizens can engage with to support their efforts to *replace plastics with plastic substitutes*:

Grenoble INP Foundation | Cellulose Valley—grenoble-inp.fr/en/academics/grenoble-inp-foundation-scholarship-program

Biodegradable Products Institute (BPI)—bpiworld.org

Apparel Impact Institute—apparelimpact.org

Players for the Planet—playersfortheplanet.org

New Earth Approved—
anewearthproject.com/collections/new-earth-approved

Notpla Disappearing Packaging—
notpla.com/sustainable-food-containers

Plastic Free July—plasticfreejuly.org

STEP 6: PHASE OUT ALL EXISTING PRODUCTION OF PLASTICS

As climate change is in the news daily, there is a need for immediate action. In North America, "wildfires are consuming three times more of the United States and Canada each year than in the 1980s, and studies predict fire and smoke to worsen."[74] Due to these wildfires, one-third of Canada and the United States experienced hazardous air quality days for several months, causing threats to human health and stresses to the environment.

As America celebrated its Independence Day in July 2023, a dangerous heat wave spread throughout one-third of the states, with 38 million Americans experiencing various heat alerts. On the macro scale, Earth's average temperature set record highs on three consecutive days, exceeding average global temperatures of 63°F by the University of Maine's Climate Reanalyzer measurements.[75] A polar explorer says global warming is evident at both poles, and "warming climates might lead to increasing risks of diseases such as the avian flu spreading in the Antarctic that will have devastating consequences for penguins and other fauna in the region."[76]

Is this a "new normal" we are settling into? A climate scientist from the University of Pennsylvania counters, "It continues to get worse. If we continue to warm the planet, we don't settle into some new state. It's an ever-moving baseline of worse and worse."[77]

After the previous steps are underway, the immediate need in Step 6 must be the universal worldwide ban on all new and existing production of virgin plastics. Stop the harm for future generations. Stop producing new plastics. From a climate perspective, the immediate need is *to phase out all production of plastics by 2030*. An existential threat of climate change drives the need to shut down new proposed plants *and* existing plastic production plants based on CO_2 and CH_4 emissions and human harm, as the hazardous impacts of the production facilities are documented. Plastic has implications for human health, the environment, and Earth's climate, as documented in Chapters 1, 4, 5, and 6. As noted throughout this book, we cannot recycle our way out of the plastics crisis! This is a reminder that plastics impact the climate, human health, and the local environment.

The obvious next step is to turn away from using plastics and to stop utilizing fossil fuels, which leads to the decision to stop producing plastics. That hard stop must be soon, as climate scientists' warnings are austere and stark.

Phasing out all existing production facilities of plastic will require the policy strength of state and federal governments. Policy tools, through governmental regulations, are available under the justification of protecting citizens' health and safety. The U.S. Department of Health and Human Services[78] and the Occupational Safety and Health Administration[79] are charged by Congress to assist citizens and workers with health and safety concerns. The U.S. Environmental Protection Agency is charged with regulating harmful chemicals in the environment.[80] Other countries offer similar governmental assistance programs to protect the well-being of their citizens. Engaging with your local elected representatives is a good start in activating the support of federal and local governmental services.

There are also numerous nonprofit organizations available to assist community working on health and safety issues. The immediate concerns of plastics pollution in the surrounding community and effects on local workers are valid starting points to organize efforts to stop plastics production and close down existing facilities. Individuals like you and I may feel overwhelmed with taking on a multibillion-dollar company. The starting seed toward any societal goal including ending plastics pollution is collective action, with individuals joining together in community-organized opposition groups.

I offer five organizing tips to consider as you work to close down oil, gas, and plastic facilities.

- Education: Your community is likely in need of education of the harms presented by plastic production facilities. Presenting educational sessions will enlarge your group of concerned citizens.

- Fact-finding: It is extremely important to work from a fact-based platform. Research the facts of your local plastics facility and its impacts on the community.

- Leadership: Seek local leaders to support your efforts. Leadership can rise from hidden corners throughout the community; be open and inclusive to all who might be impacted by the plastics facility.

- Fundraising: Funds are needed to organize an effective long-term protest that seeks to close down a multi-billion-dollar facility. Seek collaborators that have experience in fundraising.

- End goals: As you organize your community group, keep the end goal in focus and develop a strategic plan. The end goal is to permanently shut down gas, oil, and plastic production facilities.

An immediate action you can take to stop the production of plastics is to seek out where the nearest gas fracking and plastics cracking facilities are near your home. The nonprofit Halt the Harm Network offers free consultations that can help you start.[81] As you locate nearby plastic production facilities, this group will assist in organizing protest groups and public education awareness sessions of the harm presented to the local communities. Education and protest are excellent first step tools toward petitioning for the shutdown of these polluting industries.

Plastic Free Future is another community assistance nonprofit, offering outreach services and programs to create "solutions that require a shift away from the current model of excessive plastic consumption and where sustainable solutions are still a privilege. From local action and outreach to advocacy at international scales."[82]

><><><

Here is a list of experts that citizens can engage with to support their efforts to *ban all plastic production facilities*:

Halt the Harm Network—halttheharm.net

People Over Plastic—climatebase.org/company/1134410/people-over-plastic

People Over Petro—peopleoverpetro.org/project/plastic

Plastic Free Future—plastic-free-future.org

Third Act—thirdact.org

Beyond Plastics—beyondplastics.org/act

Plastic Pollution Coalition—plasticpollutioncoalition.org

THE CIRCULAR ECONOMY

"The circular economy tackles climate change and other global challenges like biodiversity loss, waste, and pollution by decoupling economic activity from the consumption of finite resources."[83]

—THE ELLEN MACARTHUR FOUNDATION

The three basic principles of the circular economy are as follows:

- Eliminate waste and pollution
- Circulate products and materials (at their highest value)
- Regenerate nature[84]

Plastics have no place in the circular economy. Plastics, all fossil-fuel-generated plastics, are not circular by their very nature.

From the extraction of gas and oil to transport, to processing and refining, to the manufacturing of plastics, to the packaging and sales of plastic products, the consumer usage, and the final disposal of most plastics (exclusive of reuse and recycling), the process is linear, not circular.

Consider the brief pathway of the reuse and recycling of plastics, and we see "downcycling." How many times is the plastic container or plastic bag reused before it is landfilled? How many cycles of recycling does plastic go through—plastic bottle recycled into carpet, computer casings, or packaging—before it is discarded? How much embedded energy is lost in discarded plastics? How soon will it be landfilled or littered into the waterways? This is not what the circular economy is meant or designed to be.

When recyclers attempt to recycle plastics, there are numerous issues and problems that surface. Friction and abrasion are involved in the mechanical belts and shakers in contact with the plastics, and when they are cleansed with a water bath, they shed microplastics.[85] A detailed Greenpeace report notes that plastic recycling processes "expose workers and adjacent communities to toxic chemicals, and globally, communities of color are more likely to suffer negative health impacts from plastic manufacturing, processing, disposal and pollution."[86]

And despite any defense of creating a better circular path for plastics, they cannot overcome their origins: plastics are fossil fuel products. By that very nature, they fail to fit the first and third primary principles of a circular economy. To "eliminate waste and pollution" and achieve the first principle of the circular economy, we must find high-quality substitutes for fossil-fuel-generated plastics. To have a "regenerative nature," the product or packaging must have a renewable source, and fossil-fuel-generated plastics simply do not.

This work toward a circular economy is essential to slowing climate change. We must take action to address the climate crisis, and material recovery is vital in that effort. The United Nations' International Resource Panel concluded that *natural resource extraction and processing* contribute to about half of all global greenhouse gas emissions. That is why the EPA is developing strategies to identify the critical actions needed to reduce the impact these materials can cause.[87]

A study published in *Nature* by twelve renowned researchers presented their findings on the intersection of plastics production and use and "carbon dioxide emissions, undermining global climate targets and the Sustainable Development Goals." After presenting four future scenarios, their final recommendation was a "bold system change [that] requires 50% reduction in future plastic demand, *complete phase-out of fossil-derived plastics*, 95% recycling rates of retrievable plastics and use of renewable energy."[88] Their conclusions would lead to a circular carbon economy that minimizes carbon dioxide emissions.

The circular economy requires upcycling material, a cascading effect of material reuse or recycling, with significant considerations toward waste reduction and rethinking product and packaging design. Plastics do not fit this circular equation, no matter what the plastics industry claims—that is simply greenwashing. Plain and simple: Plastics do not meet the definition of a circular economy and cannot be forced to fit, so we must simply move away from plastics.

Experts on *circular economy* that can offer great advice include:

Ellen MacArthur Foundation—ellenmacarthurfoundation.org/topics/circular-economy-introduction/overview

McDonough Innovation—innovation.mcdonough.com

Closed Loop Partners—closedlooppartners.com

LIFE WITHOUT PLASTICS

"If emissions are not reduced by 2030, they will need to be substantially reduced thereafter to compensate for the slow start on the path to net zero emissions. The AR6 identifies net zero CO_2 emissions as a prerequisite for halting warming at any level."

—IPCC[89]

The recognition of the planetary emergency of climate change requires the hard stop on producing fossil fuels as a necessary action, and life without plastics. The choice to migrate away from plastics is offered through a six-step bridging strategy that builds us a highway toward untangling and rethinking plastic:

- Step 1: Ban New Virgin Plastic Production Immediately
- Step 2: Reduce the Consumption of Plastics
- Step 3: Reuse—Reuse of Recovered Plastics
- Step 4: Recycle—Recycle Remaining Flows of Plastics
- Step 5: Introduce Plastic Substitutes
- Step 6: Phase Out All Existing Production of Plastics

Next we'll discuss how you can support this phase-out strategy.

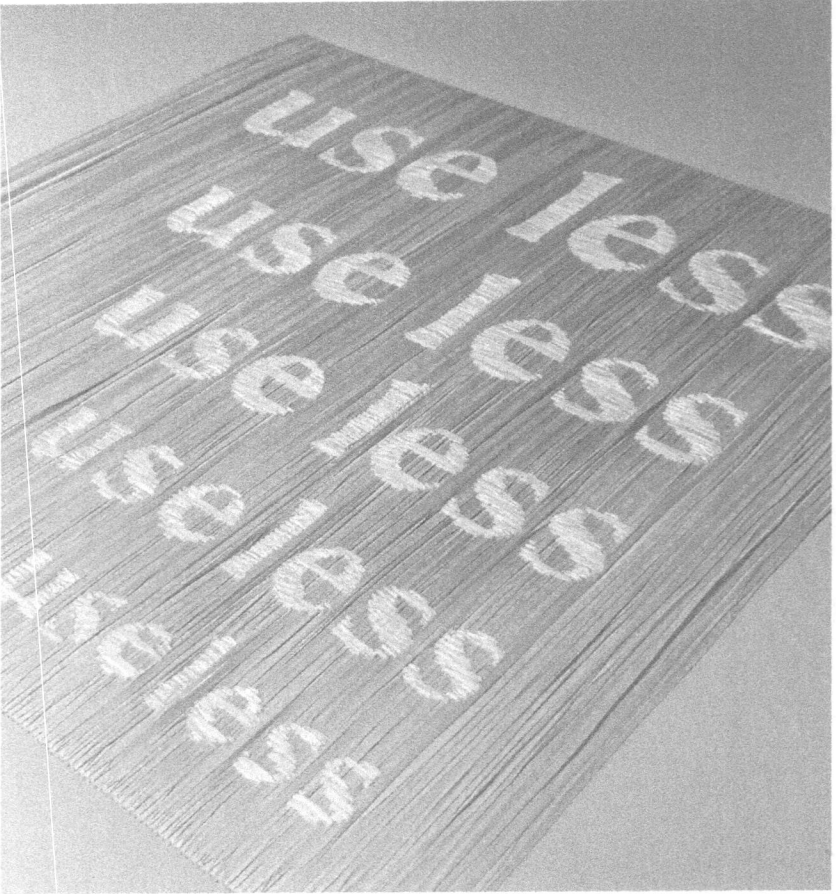

Artwork by Kalliopi Monoyios

USE LESS/USELESS

Use Less/Useless is one in a series of dental floss embroideries that asks whether Teflon, a material that has critical industrial uses but is in a class of "forever chemicals" that are environmentally problematic, is best used for thirty seconds between our teeth or as a timeless statement in art. Photo courtesy of the artist.

—KALLIOPI MONOYIOS

Chapter 10

CALL TO ACTION

"If you do nothing unexpected,
nothing unexpected happens."

—FAY WELDON[1]

As a farmer and a mechanic, Grandfather was a person of action. He sensed that waiting to do something was not good planning or leadership. He was proactive in tackling the issues presented to him. He showed me how to maintain his tractor, change the oil, tighten the bolts that vibrate loose, do a safety check, and make sure all was right with the machine, the soil, and the crops. As he got further along in age, that care was shown through family health and relationships. As a teenager, I saw the call to action through his careful movements. As a teenager, I saw the contrast between my procrastination lifestyle and his proactive approach to problem prevention. He did not take on the world's problems, but he was good at the tasks he set to work on.

The biggest threat today for farmers is climate change. It manifests as drought and harsh rainfalls, pests that live longer through the seasons, unpredictable weather patterns, and now microplastics in the soil, uptaking into our food products. The impact of these microplastics on the environment and our health is staggering. Microplastics are now found from the bottom of the deepest oceans to the top of the highest mountains. Microplastics are found in the food supply: the crops from our soil, the fish from our waters, and the plants and food products we eat. Microplastics are found in our bodies, in our organs, and in our living tissues.

When we see the enormity of the plastics problem, it's easy to become overwhelmed and paralyzed into inaction. We have been taught over the years that the plastics industry flourishes because consumers desire to utilize plastics, but did we truly ever choose this? Are we instead being fed a plastics diet beyond our will?

It is time to act. Individual, collective, governmental, and global action are needed to turn the tide against the plastics invasion. There is no time to waste. The window of opportunity to take action is now, not later. The latest IPCC report is urgent: "Mainstreaming effective and equitable climate action will not only reduce losses and damages for nature and people, it will also provide wider benefits," said IPCC Chair Hoesung Lee. "This Synthesis Report underscores the urgency of taking more ambitious action and shows that, if we act now, we can still secure a livable, sustainable future for all."[2] Let's not underestimate the power of belief in our ability to make a difference. As Gandhi believed in the transformative power of an individual, "If I keep on saying to myself that I cannot do a certain thing, it is possible that I may end by really becoming incapable of doing it. On the contrary, if I have the belief that I can do it, I shall surely acquire the capacity to do it even if I may not have it at the beginning."[3]

So let us begin with the belief that what we do matters—because it does—and look at the individual actions we can all attempt in our daily lives.

INDIVIDUAL ACTIONS— ONE GREEN STEP AT A TIME

I introduced the *One Green Step* concept on Earth Day 2010 in Austin, Texas.[4] The thought is that we can be overwhelmed by what "needs" to be done, and we might become paralyzed into inaction. Thus, taking one green step at a time, building a pattern of green steps at your pace and affordability, is the call to action. Take that first green step, maintain it for a while, then take the second step, building upon the first, and then move to the third green step. Never look backward. Keep building a platform of green steps and create a variety of "greenness" in your lifestyle.

To address the plastics crisis, please consider the following anti-plastic green step suggestions:

- Start small—**replace plastics with nonplastic equivalents**. Be creative.

- It's time for an overhaul to rid your house of plastics. Do a **room-by-room assessment** of your home's plastic use. This audit includes single-use plastics (e.g., packaging, plates, service ware, cups, bags), durable goods (e.g., lamps, rugs, office supplies, toys, hangers, furniture), and clothing (linens, towels, clothing, etc.).

- The **kitchen** deserves a great deal of attention. Remember that drinking, cooking, and microwaving with plastic containers allow microplastics to be absorbed into your body. This is a reminder to replace your nonstick cookware containing PFOA chemicals with ceramic or stainless steel cookware.

- Replace plastic bowls and plates with glass, aluminum, or ceramic. Replace plasticware with more durable nonplastic kitchenware. Plastic cutting boards transfer microplastics into your food, and some are sprayed with chemical flame retardants.

Replace those plastic cutting boards with old-fashioned wood or metal cutting board styles that have been around for centuries.

- Food safety is essential, and I highly recommend eliminating the practice of wrapping food in stretch plastic or placing it in sealable plastic bags. Most certainly, do not microwave in plastic! Instead, consider wrapping food in wax paper or aluminum foil. I have found no studies indicating the transfer of chemicals to food when wrapped in paper or aluminum. Consider reusing clean canning jars or glass containers with a sealable lid to seal food from oxygen.

- Are you a tea or coffee drinker? Check out the "carrier" of your drink. Is the tea bag made of plastic? If so, soaking it in hot water will release microplastics into your tea. Is your coffee pod made of plastic? Does that hot water stream through that coffee pod? If so, you may be at risk of drinking microplastics. Consider looking for safer alternatives. A glass carafe or ceramic pour-over cone would be safer but less convenient than a single-serving coffee pod. Use a paper coffee filter free of plastic. If you purchase coffee from a coffee shop, ask the manager to serve it in a reusable mug—you can even bring your own to some coffee shops.[5] If not reusable, at least a recyclable paper substitute would be better until reusables are more commonly acceptable.

- Buying **groceries** with the intent to avoid plastics is getting more accessible over time. Canned goods and staples can often be found in nonplastic containers. Frozen goods, cereals, pasta, cookies, and other dry goods are primarily packaged in paperboard. Dairy products are almost always packaged in plastic, but milk can be found in recyclable paperboard cartons or sometimes glass. Meat, cheeses, and other refrigerated items are commonly packaged in plastic wrap, unless you go to your local butcher where they might package in paper. Fresh produce and vegetables are not prepackaged, yet single-use plastic produce bags are conveniently available—avoid using these disposable bags and opt for reusable washable mesh produce bags instead.

- At the grocery checkout, the often-asked question is whether to bag with plastic, paper, or reusable bags. This book extensively addresses plastic grocery bags. Replace plastic bags with reusable cloth bags, and remember to wash these bags periodically. I wash them with my towels.

- The **bathroom** may be very challenging. Plastic shampoo and soap bottles can easily be replaced with safer reusable alternatives. The challenge begins with finding nonplastic toothpaste tubes and floss. Toothbrushes are also made of plastic, though some companies are offering bamboo or bioplastic. Plastic razors can be replaced with metal components, as they last longer and offer replacement razor heads. Combs and hairbrushes are most often made of plastic; however, there are plastic substitutes on the market.

- **Facial makeup** and other **skin care products** often include plasticizers. Some include glitter[6] (made of plastic) for a shimmer effect or microbeads for exfoliation. Avoid and replace any products that include plastics, as these can migrate microplastics into the environment and possibly harm your skin.

- The **bedroom** offers much room for innovation and style change. Consider replacing your home's textiles (clothing, towels, sheets, blankets, etc.) with nonplastic alternatives. Check the labels for anything that includes "poly" or is acrylic, nylon, or other synthetic materials. Note that microplastics are released through the dryer and washing machine when washing textiles. If total replacement is impossible, consider purchasing a microplastic filter for your washing machine to prevent plastics from entering the waterways.

- Seek alternative nonplastic **washing detergents**. Check the detergent label of both your laundry and dish detergents for any plastic additives. Look deep into the listed ingredients. What you might find is an ingredient called PVA, or polyvinyl alcohol. This plastic product is a thin external coating within the soap tablet (dishwasher use) or the detergent sheet (laundry machine use). Avoid these products. A quick note: Washing clothes in cold water releases fewer microplastics than washing in hot water.

- Shopping for **clothes** is often viewed as a personal style decision, not to be infringed upon with unsolicited advice. Yet much of the microplastics released into our world originate from our clothes. As you replace your clothing over time, look at the label and avoid the plastics. Migrate toward cotton, away from the "fast fashion," and toward more sustainable, timeless clothing styles.

- Replace your standard **desk office supplies** with durable, nonplastic alternatives. Consider a metal stapler rather than a plastic one, bamboo wood office organizers, refillable metal pens, wood-based pencils or metal mechanical pencils, durable metal scissors instead of ones with plastic handles, and a desk calculator void of plastics. Consider the reuse of everyday items. I use decorative coffee mugs as pen and pencil holders.

- The **living room** and **family room** offer significant opportunities for weeding out plastics. Check the labels on your furniture for polyester and other plastic fabrics. If you find plastics, donate your used furniture to a thrift store and replace it with durable, nonplastic alternatives. Gradually replace lamps, rugs, and other accessories that may contain plastics. This is an excellent opportunity to change your decor to a new style and become more Earth-friendly. Again, this may be a slow but deliberate transition.

- **Pets** are family, and they deserve a plastic-free environment too. Start with how they eat their meals—plastic food and water dishes are quick sources of microplastics. Replace them with nonplastic bowls that are dishwasher safe. Pet living environments often include blankets and cushions. Replace them with plastic-free materials. Plastic chew toys are not healthy, as pets can ingest plastics. Consider more healthy, natural replacements.

- **Toys** have changed composition over the years, and the plastics invasion has dramatically affected this sector. Sadly, children are more affected than adults by the harmful effects of plastic additives and resins, as noted in an earlier chapter. In my early childhood, most toys were constructed of tin or wood, and if broken, could be repaired. Consider replacing toys by going on a treasure hunt through antique shops. You will find safer toys for children that are reminiscent of our grandparents' stories. Consider adopting plastic toy replacement recommendations from the Play Without Plastic website (www.playwithoutplastic.com).

- Speaking of **antique shopping**, replacement clothing and furniture can be found there as alternatives to the common plastics in today's fast marketplace. Hard goods found in antique stores can be repaired and refurbished, and clothing can be retro or vintage. Kitchenware like cast iron skillets or wooden rolling pins can also be found in antique stores to restock the kitchen and replace plasticware.

- After working on the home front, consider conducting a plastics audit in other parts of your life. School, work, play, recreation, and vacation activities could benefit from a plastics audit and an evaluation of nonplastic replacements.

- As you look deeply into your single-use plastic patterns, consider **waste reduction** a better answer than replacement.

- Can you drink without a straw instead of looking for a non-plastic straw alternative?

- Can you drink that beverage in a glass, aluminum, or ceramic cup (reusable and recyclable) rather than a plastic one?

- Consider buying food condiments in glass jars rather than squeezable plastic containers.

- Purchase a refillable, reusable, nonplastic drink container that you can take with you on the go to replace the "need" for single-use plastic drink bottles.

- Create **artwork** from your discarded plastics. Engage students and adults in creatively building three-dimensional art projects that display the voice of plastic pollution, such as #TurnOffThePlasticTap by Benjamin Von Wong.[7]

- Reduce your **consumption of packaging** through smart buying practices. Bring your own bags and containers and use them to buy in bulk (by weight) when you can. Look for products packaged in paper rather than plastic, and do not accept extra packaging at checkout. As one nonprofit put it, "We did not 'ask' for our packaging to be disposable. Why should companies wait for consumers to 'ask' for reusable products and packaging?"[8]

- When flying in a commercial aircraft, **decline the plastic cup** offered for the "free" beverages. Plastic cups are not "free," considering their external environmental and human health impact. Note that some airlines are reconsidering plastic cups and replacing them with recyclable paper cups.[9]

- Consider how nonplastic **reuse** can be incorporated into your lifestyle. Repair and resale of consumer products through retail thrift positively impact environmental, economic, and social issues in communities.

- ○ Commit to washable cloth shopping bags and avoid plastic bags.

- ○ Consider your on-the-go eating preferences. Take a nonplastic reusable container to carry your leftovers home when eating out. This also works well with takeout food orders—hand your containers to the host when placing your order and have it packed without single-use utensils.

- ○ Reusable cutlery (spoons, forks, knives, etc.) on the go can replace plastic disposables and are convenient once you gain the habit.

- ○ Move to glass at home instead of relying on plastic cups or containers. Glass is reusable thousands of times, while plastics have significant environmental and health hazards.

- ○ Consider donating and purchasing your furniture through a thrift store; avoid purchasing new plastic threads. Furniture Bank, Goodwill, and Salvation Army have nationwide distribution networks and great missions.

- As you replace plastics with nonplastics, try to **recycle** the unwanted plastics at your local recycling program, depending on their collection standards. Not all plastics from your home will be recyclable, but some will be.

- ○ We cannot recycle our way out of this plastic problem, so don't justify new plastic purchases because they are recyclable.

- ○ Try to recycle what you can rather than dump it in a landfill.

- ○ If uncertain about what can be recycled locally, contact your local city, township, or county. Another resource for recycling information is Earth 911.[10]

- **Repairing** broken household items instead of quickly discarding and replacing them is part of our reduce-reuse-recycle mantra.

  ◦ Bring a **fix-it clinic** to your community.[11]

  ◦ Repairing may not be as simple as disposal, but it often is less expensive than replacement and offers longevity to our products.

  ◦ Support local repair shops. If you cannot find a repair shop, look for YouTube tutorials, ask your neighbor, or borrow a grandfather!

- **Share** your lawn mower and garden tools with your neighbors. Purchase durable, nonplastic garden equipment and create a neighborhood tool-sharing network. It's a great way to meet your neighbors and mutually support one another. There are also local tool-lending libraries to explore.[12]

- **Divest your investments from fossil fuel companies** toward greener investments. Check with a financial advisor who is well-versed in ESG investments. One recommendation is Green Century, a mutual fund company in the United States wholly owned by environmental and public health nonprofits (Disclaimer: I am not an investment advisor and make no future revenue guarantees on these recommendations.)

- **Target your retirement funds in ESG investments** to disinvest from plastics and fossil fuels. Retirement funds make up many investments into the extractive energy industry, firing up our climate. Retirement fund managers need to be instructed by their customers to invest in funds that are more focused on climate solutions and social responsibility. The SEC recently adopted rules to standardize climate-related disclosures by public companies to assist in investment decisions. Check with a financial advisor who is well-versed in ESG investments.

Gently *support family and friends* in their conversion away from plastics. Ask them to read this book and study the facts on plastics. As you discuss this with family and friends, please utilize the *no judgment* approach. It is better that we work with one another than against one another to adapt our lifestyles and address climate change.

As an individual, you have tremendous power to change your lifestyle. It begins with one green step at a time, then another, building a pattern of green steps until a series of significant actions are built upon each other. As you progress, you become a role model and mentor for friends and family. I encourage you to consider the collective actions section that follows, as collectively we can build a louder voice.

How am I getting rid of plastics in my life? One step at a time! My wife and I know the dangers of plastics to our health and the planet. Yet we find plastics throughout the house, in our kitchen, and plastics that come into the home in various ways. We are working at it one step at a time. My readers can assist, as there are areas where I've found it difficult to find plastic-free substitutes. I am looking for the following plastic-free items:

- Electric razor without a plastic shell
- Mobile cell phone without plastic components
- Laptop computer without plastic components
- Television without plastic components
- Bathroom accessories without plastic components
- Activewear shoes without plastic components

Recommendations can be sent to untanglingplastics@gmail.com

I recognize that many items in our lives will not easily convert to plastic-free alternatives. Automobiles, medical supplies, and technology tools are just a few examples. We will need our collective actions to change our channels of production and supply chains to better, more innovative ones that support our health and the health of our home planet.

COLLECTIVE ACTIONS— THROUGH COLLABORATION

"Don't wait. The time will never be just right."—**NAPOLEON HILL**[13]

Collective action begins with education and partnerships. Build new relationships with those in your community concerned about the impacts of the climate crisis. Many working to address climate emergencies are unaware of the effects of plastics on the climate and their impact on human health. Education and sharing of information will lead to collective actions. I offer the following suggestions for collection actions:

- Educate your community on **"wish recycling"** by arranging tours of the local recycling MRF and offering information on what plastics cannot be recycled.

- Educate **your community on plastics at an Earth Day** community event.[14]

- Educate yourself on the local, state, and federal legislative actions regarding plastics, and **write to your representatives for action.** Letters convey more meaning to the recipients than emails. Consider gathering friends to write a collection of letters for a more significant impact.

- **Participate in Plastics Free July**, an international effort to live free of plastics for one month of the year. The Plastics Free Foundation offers many tips for finding substitutes for everyday plastics.[15]

- **Support local plastic pollution reduction programs and increased recycling infrastructure.** Send the U.S. Conference of Mayors resolution on plastics and recycling infrastructure to your local mayor and start the conversation.[16]

- **Support the 60x40 Initiative:** earthday.org's global theme for Earth Day 2024, Planet vs. Plastics, called for the end of plastics for the sake of human and planetary health, demanding a 60 percent reduction in the production of plastics by 2040. Ask your local community to embrace this commitment.[17]

- The American Medical Association (AMA) adopted a new policy *declaring climate change a public health crisis.*[18] **Communicate this policy statement to your local health department** to gain local support and actions to address climate change as a public health crisis.

- Does your local community have a **greenhouse reduction plan or climate action plan?** Is it funded, and is the community involved? Are you involved?

- Encourage your city to **develop and implement a citywide Comprehensive Plastics Reduction Program.** (The program in Los Angeles, California, can serve as a model.[19])

- Challenge your **local government to reduce its plastic footprint.** Start by eliminating single-use plastics and expand after a few successes.

- Challenge local, regional, national, and international **companies to reduce their plastic footprint.** Write letters, speak with staff and management, and make your challenge known.

- Encourage local companies to **move toward reuse and away from disposables.** Disrupt the take-make-waste model and move to a circular model by encouraging companies to decouple their growth from plastics production and move toward reuse platforms. Work with local business leaders to set reuse and waste reduction targets.

- **Protest fracking** on public lands. Communicate support to local legislatures to protect public lands from the destructive practices of fracking . Protest local fracking efforts through grassroots mobilization and community empowerment.[20]

- Encourage your local community to **sue major plastic polluters** that contaminate your local environment. Consider **political and legal actions** to move your community toward more aggressive greenhouse reduction actions.[21]

- **Challenge corporate "carbon offsets"** to achieve greenhouse gas reductions. A carbon offset is simply a write-off and not a sincere means to address the corporation's carbon footprint. In addition, many carbon offsets are being challenged as not fully embracing the CO_2 offsets that they promise.[22] Write to your local corporate sustainability officers and CEOs and challenge the offset practice.

- If you belong to a union or environmental organization, ask the local organization to work more closely with the **BlueGreen Alliance**[23] **to support green economy jobs** in your community and eliminate the harms posed by the fossil fuel industry.

- The fossil fuel industry has historically placed its facilities in areas inhabited by Black, Brown, Indigenous, and poor communities, more commonly labeled "sacrifice zones." Support residents to **stand up against this environmental racism** and protest the construction of these facilities.

- **Write to packaging and brand companies** to change their plastic packaging to a reuse or recycling format. Consumer product companies do listen to the demands of their customers, and letters are more effective in soliciting their attention than emails. **Support companies that produce recyclable packaging** (e.g., New Earth[24] and others).

- If you are among the younger generations, **consider joining Future Coalition for Youth Climate Activists,**[25] **Gen-Z for Change,**[26] **Zero Hour,**[27] **Black Girl Environmentalist,**[28] **and Our Children's Trust.**[29] All these groups are fighting climate change on the front lines.

- If you are of the senior generation (as I am), **consider joining Third Act,**[30] founded by Bill McKibben.

- If you are a concerned mother, **join Science Moms,** an organization that supports "mothers who are concerned about their children's planet, but aren't confident in their knowledge about climate change or how they can help."[31]

- **Join the Change the Chamber network.** This organization "is a nonpartisan coalition of young adults, 100+ student groups, and other allied organizations. Within our campaigns, we educate for science-based climate action within the Chambers of Congress, trade associations (e.g., the U.S. Chamber of Commerce), corporations, government agencies, and local communities."[32]

- **Join Break Free from Plastic,** a global movement working toward "reductions in single-use plastics and to push for lasting solutions to the plastic pollution crisis." [33]

- **Join 350.org.**[34] This movement opposes the financing of fossil fuel projects and supports renewable energy projects instead.

- **Ask your local environmental groups to collectively write to USEPA** and request they finalize and implement the national USEPA Draft National Strategy to Prevent Plastic Pollution.[35]

- Promote and support a **state or federal governmental tax on the production and import of plastics for packaging**, similar to the packaging tax imposed in the EU.[36]

- **Support state legislation on collection of packaging through Extended Producer Responsibility (EPR).** The purpose of EPR legislation is to require the producer (manufacturer) of a product to be responsible for the end-of-life management of the product when the consumer is finished with it. The EPR legislation instructs the producer to establish recycling outlets for its discarded products—in other words, to avoid landfilling them. Examples are batteries, cell phones, TVs, and used tires. Plastic products can be included in EPR legislation. You can support waste reduction, reuse, and recycling contributions in EPR to optimize product design and reduce plastic waste. The National Stewardship Action Council offers model legislation and tips.[37] Plastics impact the climate across their entire life cycle, and most of these impacts occur during manufacture and supply chains, "upstream" of the consumer and waste management system.

- **Support state legislation for refillable and returnable container laws** that require the return of beverage containers (sometimes called bottle bills). The Container Recycling Institute offers model legislation and tips.[38]

- **Protest local synthetic (plastic) playing surfaces.**[39] Demonstrate to local parks and schools that petroleum-based playground surfaces are unsafe for children to play on. Offer natural solutions.

- If you own shares in a company, attend a shareholder meeting, provide a shareholder protest on **the company's plastic pollution record**, and require the company to implement a sustainability program. Research your shareholder company position through *As You Sow's* "2024 Plastic Promises Scorecard."[40]

- **Protest corporate investments in fossil fuels** and require invest-ments in renewables—"Public and private finance flows for fossil fuels are still greater than those for climate adaptation and mitiga-tion."[41] We have to put our money where our mouth is to divest from fossil fuels and invest in better options for the environment.

- Finally, **continue educating yourself and others** on these issues. I recommend research from the Intergovernmental Negotiating Committee on Plastic Pollution for **classroom study and research** on plastic pollution and its impact on the climate.[42]

NATIONAL AND INTERNATIONAL ACTIONS—FOR THE COMMON GOOD

If we continue at our current rate, by 2050 greenhouse gas emissions created by plastic production use and disposal will account for 15 per-cent of global greenhouse emissions.[43] Many systemic changes are needed to migrate away from plastics entirely, requiring national and global actions. You can be part of that action. National and interna-tional actions you can advocate for include:

- **Support the Global Plastic Treaty**—Write to your U.S. sen-ators and representatives to fully support and adopt the final global plastics resolution. Ask that the treaty address the envi-ronmental and social impacts of plastic throughout its entire life cycle. The executive director of the UN Environment Program (UNEP) requested that an Intergovernmental Negotiating Committee (INC) convene to develop "the instrument," which is to be based on a comprehensive approach that addresses the entire life cycle of plastic, including its production, design, and disposal.[44] Many in the international discussions are calling for a complete phase-out of plastics. As of the writing of this book, a final treaty has not been negotiated and approved.

- The national parks ban on single-use plastics was reversed by presidential executive order in 2025. **Petition to reinstate and expand the single-use ban to all federal properties.**

- Protect fence-line communities: **Call for suspending and denying permits** for new or expanded plastic production facilities associated with fossil fuel infrastructure projects and exports.

- **Call for the end of fossil fuel subsidies** nationally and globally. Many fracking, drilling, and plastic production facilities are constructed on public lands, public right of ways, or with public funding subsidies.

- **Write to your U.S. senators and U.S. representatives to support and fully adopt the Basel Convention, including the Plastics Amendment.** The Basel Convention, signed by 188 countries, limits or prohibits the shipment of hazardous wastes, and a 2019 amendment to the Basel Convention classified international shipments of plastic as hazardous waste.[45] The United States "has not ratified the Convention because it does not have sufficient domestic statutory authority to implement all its provisions."[46]

- **Support new federal legislation for an environmental justice energy transition plan** to migrate away from fossil fuel dependence and support a new green jobs plan for renewable energy. As we challenge the hazards of plastics, we must also challenge the hazards of reliance on fossil fuels.[47]

- **Write to world leaders at annual COP meetings to discuss climate change.** Tell world leaders, including your leaders, to eliminate plastic pollution at its source, focus on environmental justice for communities harmed by plastics production, and tax fossil fuel companies with a carbon tax.

- **Support the U.S. Plastics Pact's efforts to eliminate the production and distribution of unnecessary plastics**. Take action in your local community. View the U.S. Plastics Pact Unnecessary and Problematic Plastics website.[48]

- **Support federal legislation to ban marine plastic litter and pollution**. Explore the UN Marine Litter and Plastic Pollution Prevention Toolkit.[49]

- Encourage the federal government to use its large purchasing power to **eliminate single-use plastic items** and replace them with reusable products. Write to your legislative representative and senator to support the elimination of single-use plastic items in federal contracts.

- Encourage your legislative representative to support **federal legislation that requires eco-labels** that list the plastic content of any packaging or product.

- **Support federal legislation to ban PVC production and distribution**. PVC is used in flooring, pipes, packaging, toys, and furniture. This plastic is associated with cancer, reproductive issues, and congenital disabilities. In addition, microplastics can be found in PVC water lines.[50]

Bills in Congress you can support through your communications with your U.S. senator and house representative:

- **Break Free from Plastic Pollution Act** (118th Congress S.3127, to be reintroduced in 2025): This bill will move to amend the Solid Waste Disposal Act to reduce the production and use of certain single-use plastic products and packaging; to improve the responsibility of producers in the design, collection, reuse, recycling, and disposal of consumer products and packaging; to prevent pollution from consumer products and packaging from entering into animal and human food chains and waterways; and for other purposes.[51]

- **Green Climate Fund (GCF) Authorization Act** (118th Congress H.R.3961, to be reintroduced in 2025): This bill authorizes $4 billion in FY2024 and FY2025 for U.S. contributions to the Green Climate Fund, a fund established under the United Nations Framework Convention on Climate Change to finance projects that address climate change.[52]

- **Earth Act to Stop Climate Pollution by 2030** (118th Congress H.R. 598, to be reintroduced in 2025): This bill addresses climate change by establishing requirements concerning renewable energy, zero-emission vehicles, regenerative agriculture, and tax incentives related to climate transition costs. Specifically, the bill requires that by 2030, 100 percent of electricity sold by certain retail electric suppliers must be from renewable energy resources (e.g., wind energy); new motor vehicles (e.g., cars and trucks) sold by manufacturers must be zero-emission vehicles; and land and livestock managed by certain publicly traded corporations must be managed with regenerative agricultural practices.[53]

- **Climate Displaced Persons Act (CDPA)** (118th Congress S. 3340, to be reintroduced in 2025): This bill establishes the Global Climate Change Resilience Strategy, authorizes the admission of climate-displaced persons into the United States, and more.[54]

- **Reducing Waste in National Parks Act** (118th Congress H.R. 4561, to be reintroduced in 2025): This bill directs the National Park Service to establish a program to reduce disposable plastic products and, if applicable, eliminate the sale and distribution of disposable plastic products.[55]

- **National Container Deposit Legislation** (Bottle Bill, to be reintroduced in 2025 or 2026): This bill establishes a deposit return system for all beverage containers with a recycling refund. The draft bill would create a standardized bottle deposit system for the country and require minimum recycled content percentages in certain plastic container types.[56]

Also check for current legislation you can support by visiting Congress.gov[57] in the U.S. or the Parliament of Canada website.[58]

It's easy to get overwhelmed by what needs to be done and to become paralyzed into inaction. I've discussed a lot of actions that can be taken, but that doesn't mean you need to do it all, right now, on your own. Take a deep breath—and take one single green step. Then take another green step. Build a pattern of green steps at your own pace. Bring some friends along on the journey. Collective action in community can be energizing, encouraging, and more effective than individual efforts. At the end of the day, legislative actions on any scale are significant green steps toward eliminating the hazards of plastics in our environment.

Artwork by Kalliopi Monoyios

PARTY SIZE!

The delicate layers of a geode take hundreds of years to form, increasing in complexity as waves of mineral-laden water wash through spaces in rock. Likewise, the complexity of our entanglement with plastic has taken close to a century to form, quietly, out of sight, and is only now being revealed. Photo courtesy of the artist.

—KALLIOPI MONOYIOS

Epilogue

DO NO HARM

"With malice toward none and charity to all."

—ABRAHAM LINCOLN[1]

'd like to begin and end with a visit to my grandfather. He taught me the farmer's ethic that nature's growth depends on the previous generation's nurture of the environment and our care to do no harm. He also taught me to care for Earth as our home. This is a primary tenet of the precautionary principle, first presented in Chapter 1, to establish protection against the harmful effects of plastics on humans and Earth, our common home.

The precautionary principle encourages protecting human health and the environment's safety, where the risks may be difficult to identify or determine with certainty. Two current examples in today's world include the responsibilities of physicians and the role of governmental consumer safeguards. The physician must first do no harm in treating a sick person. The government's responsibility is to develop regulatory decisions using the precautionary principle, as written in the safety code framework, to protect its citizens' health and avoid harm. The precautionary principle provides the framework for public safety and public health.

In the application of the precautionary principle, I propose five specific action values regarding plastics in our environment:

- *Prevention of harm*: taking preventive action in the face of uncertainty (e.g., avoidance of CO_2 releases into the environment). This prevention is best achieved through the *reduction and avoidance of the use of plastics.*

- *Burden of proof*: shifting the burden of proof of harm to the advocates of an activity (e.g., production of fossil fuel products); *the burden of proof that it is a safe product should be shifted to the producer.* The consumer should not be placed with the burden of proving harm.

- *Innovation of alternatives*: exploring a wide range of alternatives to harmful actions through innovation, such as *creating new replacement products* through governmental grants and business innovation centers.

- *Intentionality*: exploring—with a whole intent—to create a new world where humans *live in closer harmony with the environment.*

- *Inclusion: involving all affected by the decision-making processes,* including the disenfranchised poor and those whose health is affected by the fossil-fuel and plastic-making industries.

In applying these specific proactive values, human health is the primary beneficiary, with secondary benefits to the environment and the climate. By shifting the means of production from an adverse defensive position toward a consumer-friendly base, product innovation can flourish. Through the addition of intentionality, product design can respect the interconnectedness of human needs and the natural environment. Through consumer focus groups that discuss product safety and desire to avoid harm from the proposed products, producers can begin to embrace inclusion in the decision-making process.

The precautionary principle, through the thoughtful application of these five action values, must be applied to global plastics pollution. As noted in prior chapters, plastics directly harm the climate, human health, and the local environment. Applying the do-no-harm precautionary principle would require following the pathway toward eliminating plastics and fully replacing plastics with safe substitutes.

"See to it that you do not spoil and destroy My world;
 for if you do, there will be no one else to repair it."

—MIDRASH ECCLESIASTES RABBAH 7:13[2]

RESTORATIVE AND REGENERATIVE

"I did then what I knew how to do.
Now that I know better, I do better."

—MAYA ANGELOU

To address the damages of climate change on Earth, our common home, we collectively need to focus on a multipronged approach that eliminates the use of fossil fuels. We know now that methane and carbon dioxide are released from the extraction, processing, and use of fossil fuel products. A lot of attention is focused on transportation and electricity uses of fossil fuels and finding alternative renewable energy sources that can replace fossil fuels. Transformation of the transportation and electricity sectors is necessary. However, often forgotten in this conversion are the millions of consumers and industrial products manufactured with fossil fuel sources. Restoration of Earth requires a full stop on all products made through fossil fuel sources, including the production of plastic products.

Planting the seed for regenerative growth in an age beyond plastics requires a full commitment to the restoration of Earth, the elimination of fossil fuel extraction and processing, and a full stop to the production of plastics. Regenerative growth begins with the innovation and use of plastic substitutes that are environmentally safe and nontoxic to human health.

Jamais Cascio spoke memorably on the need to work toward more than just sustainability, but toward "an environment and a civilization able to handle unexpected changes without threatening to collapse; . . . it would be regenerative and diverse, relying on the capacity to absorb shocks like the popped housing bubble or rising sea levels and evolve with them. In a word, it would be resilient."[3]

To achieve the sort of resilience and regenerative world Cascio speaks of, we need Restorative Environmental Justice, which "means that all communities have the right to clean, healthy, and safe places to live, work and play."[4] Dr. Ana Baptista, professor at the Milano School of Policy, Management & Environment and co-director of the Tishman Environment & Design Center at The New School, and Jose Bravo, executive director of the Just Transition Alliance, explain that though they live under constant threats, Environmental Justice communities can "teach us about what it means to be resourceful, resilient and take risks in these trying times. The will to not just [survive]—but to thrive and flourish—will not be diminished."[5]

Restorative Environmental Justice requires a full stop to the use of all fossil fuel sources, including the production of plastics. As Dr. Ana Baptista advises, we can learn from Environmental Justice communities to work toward sustainability and resilience. However, complete regenerative growth beyond plastics relies on our full commitment to the restoration of Earth, the elimination of fossil fuel extraction and processing, and a phasing out leading to the elimination of plastics. Regenerative growth requires the innovation and use of plastic substitutes that are environmentally safe and nontoxic to human health.

INTENTIONALITY

"In a time of destruction, create something."

—MAXINE HONG KINGSTON[6]

Elizabeth Kolbert notes in her recent book *The Sixth Extinction: An Unnatural History*, "There have been five mass extinctions, when the diversity of life on earth suddenly and dramatically contracted. Scientists around the world are currently monitoring the sixth extinction, predicted to be the most devastating extinction event since the asteroid impact that wiped out the dinosaurs. This time around, the cataclysm is us."[7]

We are now in the midst of the sixth extinction in geological time. As the climate worsens, so do our prospects of survival. Strong action is necessary now to prevent this cataclysmic extinction event looming on the horizon.

We must be *intentional* in our actions to correct the harm exhibited by plastics.

Intentionality is at the heart of human-induced progress—whether we apply the value of "supportive" or "harmful" progress in an opinioned and debatable matter in each situation.[8] We need human imagination, human intentionality, and human innovation, with critical urgency, to protect our people, environment, and planet.

"Wealth consists not in having great possessions,
 but in having few wants."**—EPICTETUS**

TRANSFORMATIVE CHANGE

"The Stone Age did not end because humans ran out of stones.
It ended because it was time for a re-think about how we live."

—WILLIAM MCDONOUGH[9]

It is now time for a transformative change in how humans "use" Earth's resources. Since the start of the industrial age, drilling and excavating has been the source of fossil fuel "resources" that generated our world economies, not to mention heating our homes and providing our means of transportation. Humans for a long time hardly considered as part of this equation the externalities of the environmental and human costs of fossil fuel extraction and use. Plastic creation and use, as well as disposal abuse, illustrate the consequences of "using" the resources of Earth in a destructive manner.

Migrating to a more constructive lifestyle requires a transformative change in how we view our human relationship with Earth—toward coexistence and interdependence rather than dominance and destruction. A recent IPCC report advises, "Limiting warming to 1.5°C [2.7°F] above pre-industrial levels would require *transformative systemic change*, integrated with sustainable development. Such change would require the upscaling and acceleration of implementing far-reaching, multilevel, and cross-sectoral climate mitigation and addressing barriers."[10] To meet this global challenge, the IPCC is advising us to work together, across cultures and nations, to address climate mitigation and its barriers.

In his recent book *How to Know a Person*, David Brooks shares the story of Ludwig Gutman, a Jewish doctor who escaped to England from Germany during World War II. While working in a hospital serving people with paraplegia, Gutman refused to accept that a spinal injury was a death sentence and started getting his patients out of bed and forcing them into some form of activity.

When challenged by his colleagues at the hospital accustomed to merely managing the decline of their patients, one asked, "Who do you think they are?" Gutman replied, "They are the best of men." Gutman's continued efforts to organize games for the injured men eventually led to the creation of the Paralympics.[11] What Gutman practiced is a powerful example of seeing the potential in others and treating them with respect, which led to *transformative change in our thinking patterns and lifestyles*.

This book is about envisioning and working toward a plastic-free future, which can lead to a *transformational change in lifestyle*. Eliminating the production of plastics is a *systemic, transformative change*. Encouraging the production of alternatives to plastics can be an *industrial transformative change*. Migrating away from plastics and seeking safe alternatives is a *personal transformative change*.

Consumers can support this transformation by assessing their current levels of consumerism (likely overconsumption) and working toward waste reduction before purchasing more sustainable replacements for plastic products. The plastics industry has moved the entire consumer world into a take-make-waste system of disposals. We consumers need to take a step back, look at our excessive purchases, and ask first if there is a need, and second if there is an alternative to plastics. Economist Juliet Schor "argues for a new kind of analysis: looking at the costs of consumption in terms of the environment and our quality of life."[12] I would add to the equation the cost of consumerism contributing to climate change.

The business community can support this transformation by embracing "ambitious science-based actions and adopting nature-supportive pathways that align with the mitigation hierarchy avoid-reduce-restore-transform."[13] This pathway leads toward waste reduction, zero waste, circular economy structures, and offering nonplastic products that are safe for humans and the environment.

Local governments can support transformative change by challenging the status quo. An example is Ford County, Kansas, which is suing the American Chemistry Council, Eastman, ExxonMobil, and LyondellBasell, claiming that they lied about their promotions of plastic recycling efforts. In the lawsuit, on behalf of its citizens, Ford County claims "the companies made profit-driven decisions for decades to deceive consumers about plastics recycling's effectiveness" without improving actual recycling services. The class action lawsuit is founded on the basis of deceptive marketing claims.[14]

There have also been discussions of transformative change in plastics production and use on the global level. At the writing of this book, the world nations are in the fifth set of discussions toward creating the Global Plastics Treaty, a "legally binding instrument" on plastic pollution. The Intergovernmental Negotiating Committee (INC) was established by a United Nations Environment Assembly (UNEA) resolution entitled "End Plastic Pollution: Towards an International Legally Binding Instrument." The INC is charged to develop "an international legally binding instrument on plastic pollution, including in the marine environment . . . based on a comprehensive approach that addresses the full life cycle of plastic, taking into account among other things, the principles of the Rio Declaration on Environment and Development, as well as national circumstances and capabilities."[15]

These 175 UNEA members have been meeting periodically since 2022 to create this global treaty on plastics. The nations are split in the discussions, with at least ninety-four nations within a "High Ambition Coalition" while a couple dozen other nations are partnering with the oil industry as the "Like Minded Group."[16]

The latest discussions ended poorly, as the fossil fuel industry influenced and disrupted the decision-making process.[17] The key issue discussed was a cap on global plastic production, yet lobbyists from plastic producers were "furiously arguing against any attempt to restrain the amount that can be produced."[18] There will be another set of discussions set for late 2025, yet new ground rules need to be established to limit interference from the plastic pollution industries.

If this Global Plastics Treaty is eventually approved, I encourage all readers to petition their national leaders for ratification of this treaty without modifications or exceptions. In the United States, write to your U.S. senators and congressional representatives. In Canada, write to your prime minister, senators, and the members of the House of Commons.

POST-PLASTICS AGE

"But if we break down the walls that hem us in, if we step out into the open and dare to embrace new beginnings, everything is possible."

—ANGELA MERKEL[19]

These words, delivered by German Chancellor Angela Merkel at a Harvard commencement, were intended to speak to the fragility of democracy worldwide.[20] However, these exact words and sentiments apply here as we talk of turning the page away from the Universal Plastics Age to a new Post-Plastics Age in which humans respect Earth and do not harm the climate. As Merkel states, we need to "break down the walls that hem us in."

In Chapter 2, I introduced three plastic ages. The Wartime Plastics Age spanned the 1920s through the 1940s, when plastics were mainly in the discovery and development stage in scientific labs. In this era, necessity was the driver of invention. The Jetsons Plastics Age began in post-war 1946 and continued throughout the next two decades, with the focus on the consumer. This era saw the start of plastics becoming a part of everyday consumer goods, from kitchen appliances to furniture. The Universal Plastics Age began in the 1960s with the introduction of single-use disposable plastics and gained much steam in the mid-'70s with the advent of the single-use plastic bag. Other symbols of the era include single-serving plastics beverage containers and disposable plastic straws—heralds of Americans' new mindset of consumption and disposal over sustainable reuse.

We still consume large quantities of plastics today as the Universal Plastics Age continues on, and plastics have become among the most ubiquitous artificial materials on Earth. I believe it is time to leave history behind and progress to a fourth era: the Post-Plastics Age.

How and when will we usher in the Post-Plastics Age? Removing our reliance on plastics will require a transformative change. It will take action from individuals and policymakers to break our addiction to plastics and move forward toward a healthier, more sustainable future. That is why I propose the six-step policy program in Chapter 9 and the call-to-action steps in Chapter 10.

Transformative change does not happen on its own; it takes people to make it happen. It requires innovation and intentionality, hope and action, to spark change from ourselves in our own lives, to our communities, to our nations, to our global home, Planet Earth. The great plastic awakening must happen now.

HOPE AND ACT

"There is hope—I've seen it—but it does not come from the governments or corporations. It comes from the people."

—GRETA THUNBERG[21]

"We can no longer let the people in power decide what hope is. Hope is not passive. Hope is not blah blah blah. Hope is telling the truth. Hope is taking action."

—GRETA THUNBERG[22]

Hope for the future requires looking through a different set of lenses. Hope is a feeling of expectation for something to happen. Hope is intrinsically linked to action. As Greta Thunberg astutely notes, hope is taking action. To trust in a better future requires hope *and* action.

Hope also requires collective action—working together toward a common goal. Reaching for a fossil fuel-free and plastic-free world will require transformative change.

Hope is in the wind. I see hope in the college students I teach. They yearn for a future of green energy, clean water, safe food, and healthy lives. I also see hope in the international climate talks. At the COP28 UN climate talks in December 2023, 195 countries agreed to transition away "from planet-warming fossil fuels—the first time they've made that crucial pledge in decades of UN climate talks."[23]

I hope you will embrace this book's message and act on it. We need to kick the habit of plastic consumption. It is time to take action now! Remember Ziggy's famous advice about changing how you see a hopeless situation with a turn in the direction you are facing.

Reduce your plastic footprint; walk lightly on Earth, our common home.

The great plastic awakening must happen now!

Readers may contact me at: untanglingplastics@gmail.com

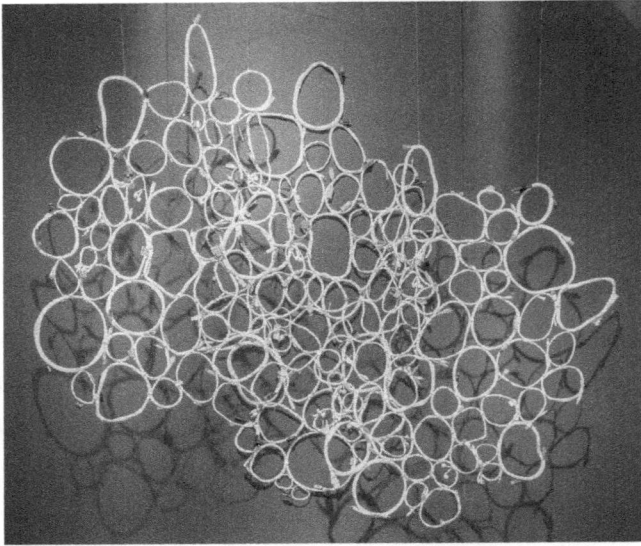

Artwork by Kalliopi Monoyios. Photo by Wes Magyar.

ACCUMULATED DEPRECIATION

Accumulated Depreciation is an accounting concept that quantifies the amount of value a given asset loses each year. Interestingly, this loss in value is considered an asset, not a liability. My installation by the same name is composed of discarded cords which are, in most cases, perfectly functional but have been rendered useless as people follow corporations' cues to upgrade their tech. Alone, these discarded cords are worthless, but collected together and reinvented as art, we recover some of their original value. The wrapped coils are arranged together in clusters of cells that feel as though they are dividing and growing, almost taking on a life of their own. Depending on the size of the installation, this can be experienced as threatening, or even metastasizing, like our unchecked consumption. Photo by Wes Magyar.

—KALLIOPI MONOYIOS

Appendix

REFLECTIONS

"Future generations will never forgive us if we miss the opportunity to protect our common home. We have inherited a garden; we must not leave a desert for our children."

—Joint statement from world religious leaders[1]

"When we touch the earth mindfully every step will bring peace and joy to the world."

—THICH NHAT HANH[2]

THE EARTH CHARTER[3]

We stand at a critical moment in Earth's history, a time when humanity must choose its future. As the world becomes increasingly interdependent and fragile, the future at once holds great peril and great promise. To move forward we must recognize that in the midst of a magnificent diversity of cultures and life forms, we are one human family and one Earth community with a common destiny. We must join together to bring forth a sustainable global society founded on respect for nature, universal human rights, economic justice, and a culture of peace. Toward this end, it is imperative that we, the peoples of Earth, declare our responsibility to one another, to the greater community of life, and to future generations.

Earth, Our Home

Humanity is part of a vast evolving universe. Earth, our home, is alive with a unique community of life. The forces of nature make existence a demanding and uncertain adventure, but Earth has provided the conditions essential to life's evolution. The resilience of the community of life and the well-being of humanity depend upon preserving a healthy biosphere with all its ecological systems, a rich variety of plants and animals, fertile soils, pure waters, and clean air. The global environment with its finite resources is a common concern of all peoples. The protection of Earth's vitality, diversity, and beauty is a sacred trust.

The Global Situation

The dominant patterns of production and consumption are causing environmental devastation, the depletion of resources, and a massive extinction of species. Communities are being undermined. The benefits of development are not shared equitably and the gap between rich and poor is widening. Injustice, poverty, ignorance, and violent conflict are widespread and the cause of great suffering. An unprecedented rise in human population has overburdened ecological and social systems. The foundations of global security are threatened. These trends are perilous—but not inevitable.

The Challenges Ahead

The choice is ours: form a global partnership to care for Earth and one another or risk the destruction of ourselves and the diversity of life. Fundamental changes are needed in our values, institutions, and ways of living. We must realize that when basic needs have been met, human development is primarily about being more, not having more. We have the knowledge and technology to provide for all and to reduce our impacts on the environment. The emergence of a global civil society is creating new opportunities to build a democratic and humane world. Our environmental, economic, political, social, and spiritual challenges are interconnected, and together we can forge inclusive solutions.

Universal Responsibility

To realize these aspirations, we must decide to live with a sense of universal responsibility, identifying ourselves with the whole Earth community as well as our local communities. We are at once citizens of different nations and of one world in which the local and global are linked. Everyone shares responsibility for the present and future well-being of the human family and the larger living world. The spirit of human solidarity and kinship with all life is strengthened when we live with reverence for the mystery of being, gratitude for the gift of life, and humility regarding the human place in nature.

We urgently need a shared vision of basic values to provide an ethical foundation for the emerging world community. Therefore, together in hope we affirm the following interdependent principles for a sustainable way of life as a common standard by which the conduct of all individuals, organizations, businesses, governments, and transnational institutions is to be guided and assessed.

BRIDGE TO BUSAN: DECLARATION ON PRIMARY PLASTIC POLYMERS[4]

We, the undersigned members of the intergovernmental negotiating committee (INC) and those concerned about the many harms of plastic pollution to human health and the environment, are committed to ending plastic pollution worldwide.

We reaffirm the mandate of United Nations Environment Assembly (UNEA) Resolution 5/14 to develop an international legally binding instrument on plastic pollution, including in the marine environment, based on a comprehensive approach that addresses the full lifecycle of plastics.

We emphasize that the full lifecycle of plastics includes the production of primary plastic polymers.

Studies show that the world cannot achieve its goals of ending plastic pollution and limiting global average temperature rise to less than 1.5° Celsius if the unsustainable production of primary plastic polymers is not addressed.

Left unaddressed, production of primary plastic polymers is projected to increase exponentially through 2050 and could overwhelm national waste management and recycling programs, even after significant improvements supported by the new instrument.

Addressing the unsustainable production of primary plastic polymers is not only essential to ending plastic pollution worldwide; it also represents one of the most efficient and cost-effective approaches to managing the plastic pollution problem.

Moreover, a balancing of efforts across the full lifecycle of plastics—from production and design through waste management and remediation—is necessary to equitably distribute the overall burden of efforts shared among countries, each of which must contribute to achieving the collective goals of the new instrument.

For these reasons, we call on members to:

COMMIT to achieve sustainable levels of production of primary plastic polymers. This includes ensuring production matches ambitions for a circular economy for plastics, while aligning with the Paris Agreement goal of limiting warming to 1.5°C.

ENSURE transparency in the production of primary plastic polymers. This includes reporting of data on the production of primary plastic polymers to close information gaps, assess progress, and inform priorities.

AGREE to a global objective regarding the sustainable production of primary plastic polymers. This may include production freezes at specified levels, production reductions against agreed baselines, or other agreed constraints to prevent the unsustainable production of primary plastic polymers.

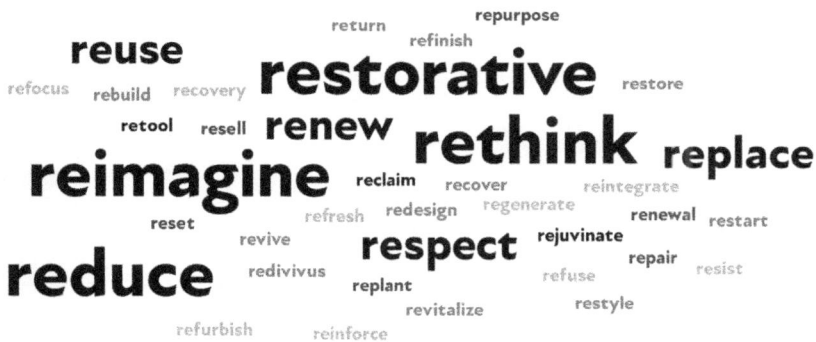

return repurpose
reuse refinish
refocus rebuild recovery **restorative** restore
retool resell renew **rethink** replace
reimagine reclaim recover reintegrate
reset refresh redesign regenerate renewal restart
revive **respect** rejuvinate
reduce redivivus repair resist
replant refuse
revitalize restyle
refurbish reinforce

In memory of my grandfather, please reduce your plastic footprint and walk lightly on our precious home, Mother Earth.

NOTES

PREFACE

1. International Panel on Climate Change (IPCC). (n.d.). *AR6 Synthesis Report: Headline Statements*. Retrieved June 16, 2025, from https://www.ipcc.ch/report/ar6/syr/resources/spm-headline-statements/.

INTRODUCTION

1. Center for Climate Integrity (CCI). (2024, February). *The Fraud of Plastic Recycling*. https://climateintegrity.org/uploads/media/Fraud-of-Plastic-Recycling-2024.pdf.

CHAPTER 1

1. Fuller, R. B. (1975). *Synergetics*. Macmillan.
2. Zero Waste International Alliance. (2008). *Zero-Waste Definition*. ZWIA.org. http://zwia.org/zero-waste-definition/; United Nations. (1992). *Rio Declaration on Environment and Development*. https://www.un.org/en/development/desa/population/migration/generalassembly/docs/globalcompact/A_CONF.151_26_Vol.I_Declaration.pdf.
3. Covey, S. R. (1990). *The 7 Habits of Highly Effective People*. Simon & Schuster.
4. Suzuki, D. W., & Hanington, I. (2018, May 3). Cutting Through Polluted Public Discourse. *Science Matters*. https://davidsuzuki.org/story/cutting-through-polluted-public-discourse/.

5. Intergovernmental Panel on Climate Change (IPCC). (2018). *Global Warming of 1.5°C*. United Nations. https://www.ipcc.ch/sr15/.

6. Climate Central. (2025, January 22). *Fastest Warming Seasons*. https://www.climatecentral.org/climate-matters/fastest-warming-seasons-2025.

7. United Nations. (1992). *Rio Declaration on Environment and Development*. https://www.un.org/en/development/desa/population/migration/generalassembly/docs/globalcompact/A_CONF.151_26_Vol.I_Declaration.pdf.

8. Oreskes, N., Lanier-Christensen, C., Conway, H., & Macfarlane, A. (2024). Climate Change and the Clean Air Act of 1970 Part I: The Scientific Basis. *Ecology Law Quarterly, 50* (3). https://doi.org/10.15779/Z38DV1CP99.

9. Yoder, K. (2024, August 5). The Lost History of What Americans Knew about Climate Change in the 1960s. *Grist*. https://grist.org/science/lost-history-climate-1960s-clean-air-act-supreme-court/.

10. Broecker, W. S. (1975). Climate Change: Are We on the Brink of a Pronounced Global Warming? *Science, 189* (4201), 460–463. https://www.science.org/doi/10.1126/science.189.4201.460.

11. Kennedy, J. F. (1961, May 25). *Excerpt from an Address Before a Joint Session of Congress*. JFK Presidential Library and Museum. https://www.jfklibrary.org/learn/about-jfk/historic-speeches/address-to-joint-session-of-congress-may-25-1961.

12. VELCRO. (n.d.). *Securing Success for NASA Astronauts*. Retrieved June 11, 2025, from https://www.velcro.com/original-thinking/securing-success-for-nasa-astronauts/.

13. Pierre, J. (2023, February 7). Intentionality. In *Stanford Encyclopedia of Philosophy*. https://plato.stanford.edu/entries/intentionality/.

14. Pinto-Bazurco, J. F. (2020, October 23). *The Precautionary Principle*. International Institute for Sustainable Development. https://www.iisd.org/articles/precautionary-principle.

15. United Nations. (1992). *Rio Declaration on Environment and Development*. https://www.un.org/en/development/desa/population/migration/generalassembly/docs/globalcompact/A_CONF.151_26_Vol.I_Declaration.pdf.

CHAPTER 2

1. Commoner, B. (1971). *The Closing Circle: Nature, Man, and Technology*. Dover Publications.

2. Nichols, M. (Director). (1967). *The Graduate* [Film]. United Artists.

3. Dow Chemical. (n.d.). *Financial Reporting, Dow Investor Relations*. Retrieved January 23, 2021, from https://investors.dow.com/en/financial-reporting/default.aspx#earnings.

4. etymonline. (n.d.). Plastic. In *Online Etymology Dictionary*. Retrieved June 11, 2025, from https://www.etymonline.com/search?q=plastic.

5. Science History Institute. (n.d.). *Science of Plastics*. Retrieved June 11, 2025, from https://www.sciencehistory.org/science-of-plastics.

6. Ibid.

7. Ibid.

8. Freinkel, S. (2011). *Plastic: A Toxic Love Story*. Houghton Mifflin Harcourt.

9. Ibid.

10. Ibid.

11. *Time*. (1924, September 22). *Leo H. Baekeland | Sep. 22, 1924* [magazine cover]. http://content.time.com/time/covers/0,16641,19240922,00.html.

12. Dupont. (1988). *Nylon: A DuPont Invention*. Retrieved June 11, 2025, from https://digital.hagley.org/pam_99_008#page/2/mode/2up.

13. Wolfe, A. J. (2008, October 3). Nylon: A Revolution in Textiles. *Distillations Magazine*. https://www.sciencehistory.org/stories/magazine/nylon-a-revolution-in-textiles/.

14. Hendrickson, K. E. (2014). *The Encyclopedia of the Industrial Revolution in World History*. Rowman & Littlefield.

15. Science History Institute. (n.d.). *Science of Plastics*. Retrieved June 11, 2025, from https://www.sciencehistory.org/science-of-plastics.

16. Leighton, J. L. (1942, August). Plastics Come of Age. *Harper's Magazine*, p. 306.

17. Freinkel, S. (2011). *Plastic: A Toxic Love Story*. Houghton Mifflin Harcourt.

18. Ibid.

19. Severson, A. (2010, March 25). Pontiac Fiero: The Definitive History. *Jalopnik*. https://www.jalopnik.com/pontiac-fiero-the-definitive-history-5501545/.

20. Leigh, E. (2011, March 17). The History of Plastic Bottles. *Recycle Nation*. https://recyclenation.com/2011/03/history-plastic-bottles-recycle/.

21. Pandal, N. (2018, August 10). Birth of the Bottled Water Industry. *BCC Research*. https://blog.bccresearch.com/birth-of-the-bottled-water-industry.

22. Taylor, S. L. (2017, June 28). A Million Bottles a Minute: World's Plastic Binge "as Dangerous as Climate Change." *The Guardian*. https://www.theguardian.com/environment/2017/jun/28/a-million-a-minute-worlds-plastic-bottle-binge-as-dangerous-as-climate-change.

23. Collins, S. (2020). *Plastic Facts & Statistics*. Container Recycling Institute. https://www.container-recycling.org/index.php/factsstatistics/plastic.

24. Gibbens, S. (2019, January 2). A Brief History of How Plastic Straws Took Over the World. *National Geographic.* https://www.nationalgeographic.com/environment/article/news-plastic-drinking-straw-history-ban.

25. Ibid.

26. Impasta Straws. (2022, October 3). *Which States Have Banned Plastic Straws?* https://impastastraws.com/blogs/plastic-straw-bans-101/which-states-in-us-have-banned-plastic-straws.

27. BBC. (2020, October 1). Plastic Straw Ban in England Comes into Force. *BBC News.* https://www.bbc.com/news/uk-england-54366461.

28. Roland, G., Jambeck, J. R., & Law, K. L. (2017, July 19). Production, Use, and Fate of All Plastics Ever Made. *Science Advances, 3*(7). https://doi.org/10.1126/sciadv.1700782.

29. Shirley, S. (2018, September 13*). The History of The Shopping Bag: Part Two—Single-Use Plastic Bags.* Factory Direct Promos.com. https://www.factorydirectpromos.com/blog/the-history-of-single-use-plastic-bags/.

30. Plastic Oceans. (2021). *Plastic Pollution Facts.* https://justoneocean.org/ocean-plastic-facts.

31. Ibid.

32. Jenner, O. (2017, April 5). The True Costs of Single-Use Plastic Bags. *Sierra Club Maine Chapter.* https://www.sierraclub.org/maine/blog/2017/04/true-costs-single-use-plastic-bags.

33. United Nations. (2018, December 6). *Legal Limits on Single-Use Plastics and Microplastics.* https://www.unep.org/resources/report/legal-limits-single-use-plastics-and-microplastics.

34. National Conference of State Legislatures. (2021, February 8). *State Plastic Bag Legislation.* https://www.ncsl.org/environment-and-natural-resources/state-plastic-bag-legislation.

35. Surfrider Foundation. (2019, October 10). *The Latest Plastic Bag Laws and Map.* https://www.surfrider.org/coastal-blog/entry/the-latest-plastic-bag-laws-and-maps.

36. Chappell, B. (2024, September 25). California's First Plastic Bag Ban Made Things Worse. Now It's Trying Again. *NPR.* https://www.npr.org/people/14562108/bill-chappell.

37. Surfrider Foundation. (2019, October 10). *The Latest Plastic Bag Laws and Map.* https://www.surfrider.org/coastal-blog/entry/the-latest-plastic-bag-laws-and-maps.

38. Coda. (2018, May 4). *World Cup: How Plastic Revolutionized the Modern Football.* https://www.coda-plastics.co.uk/blog/world-cup-how-plastic-revolutionised-the-modern-football.

39. Wham-O. (2015). *The History of Wham-O.* https://web.archive.org/web/20150517074123/http://www.wham-o.com/history.html.

40. Yeaworth, I. & Doughton, R. (Directors). (1958). *The Blob* [Film]. Paramount Pictures.

41. Carrington, D. (2025, February 22). "Technofossils": How Humanity's Eternal Testament Will Be Plastic Bags, Cheap Clothes and Chicken Bones. *The Guardian.* https://www.theguardian.com/science/2025/feb/22/technofossils-how-plastic-bags-and-chicken-bones-will-become-our-eternal-legacy.

42. Miller, C. (2024, March 1). *The Realities of Plastics Recycling.* Waste360. https://www.waste360.com/plastics/the-realities-of-plastics-recycling.

43. Parker, L. (2018, December 20). A Whopping 91 Percent of Plastic Isn't Recycled. *National Geographic.* https://education.nationalgeographic.org/resource/whopping-91-percent-plastic-isnt-recycled/.

44. Freinkel, S. (2011). *Plastic: A Toxic Love Story.* Houghton Mifflin Harcourt.

45. Ghaddar, A., & Bousso R. (2018, October 4). Rising Use of Plastics to Drive Oil Demand to 2050: IEA. *Reuters.* https://www.reuters.com/article/us-petrochemicals-iea/rising-use-of-plastics-to-drive-oil-demand-to-2050-iea-idUSKCN1ME2QD.

46. Tiseo, I. (2021, January 27). *Global Market Value of Plastic 2018–2027.* Statista. https://www.statista.com/statistics/1060583/global-market-value-of-plastic/#statisticContainer.

CHAPTER 3

1. Science History Institute. (n.d.). *Science of Plastics.* Retrieved June 11, 2025, from https://www.sciencehistory.org/science-of-plastics.

2. Polyethylene. (2025, May 23). In *Wikipedia.* https://en.wikipedia.org/wiki/Polyethylene.

3. Science History Institute. (n.d.). *Science of Plastics.* Retrieved June 11, 2025. from https://www.sciencehistory.org/science-of-plastics.

4. United States Environmental Protection Agency. (2024, November 21). *Plastics: Material-Specific Data.* https://www.epa.gov/facts-and-figures-about-materials-waste-and-recycling/plastics-material-specific-data.

5. Sullivan, L. (2020, September 11). How Big Oil Misled the Public Into Believing Plastic Would Be Recycled. *NPR.* https://www.npr.org/2020/09/11/897692090/how-big-oil-misled-the-public-into-believing-plastic-would-be-recycled.

6. Ibid.

7. Ibid.

8. American Chemistry Council. (n.d.). *Plastics.* Retrieved June 11, 2025, from https://www.americanchemistry.com/chemistry-in-america/chemistry-in-everyday-products/plastics/.

9.   Mhatre, A. (2018, August 5). Piling Up: Drowning in a Sea of Plastic. *CBS News*. https://www.cbsnews.com/news/piling-up-drowning-in-a-sea-of-plastic/.

10.   Plastic containers labeled with code for basic material used in bottle or container, Ohio Revised Code, Title 37, Section 3734.60 (1989). https://codes.ohio.gov/ohio-revised-code/section-3734.60.

11.   Advancing Standards Transforming Markets (ASTM). (2022, January 4). *Standard Practice for Coding Plastic Manufactured Articles for Resin Identification*. https://store.astm.org/d7611_d7611m-21.html.

12.   Yoder, K. (2024, June 12). How the Recycling Symbol Lost Its Meaning. *Grist*. https://grist.org/culture/recycling-symbol-logo-plastic-design.

13.   Bruggers, J. (2024, April 1). Kraft Heinz Faces Shareholder Vote on Its "Deceptive" Recycling Labels. *Inside Climate News*. https://insideclimatenews.org/news/01042024/kraft-heinz-recycling-labels.

14.   Shen, Y. (2023, March 21). Resin Identification Codes in the United States: A Practical Guide. *ComplianceGate*. https://www.compliancegate.com/resin-identification-codes/.

15.   Environmental advertising: recycling symbol: recyclability: products and packaging, California Senate Bill No. 343, Ch. 507. (2021). https://leginfo.legislature.ca.gov/faces/billTextClient.xhtml?bill_id=202120220SB343.

CHAPTER 4

1.   Law, K.L., Starr, N., Siegler, T. R., Jambeck, J. R., Mallos, N. J., & Leonard, G. H. (2020). The United States' Contribution of Plastic Waste to Land and Ocean. *Science Advances, 6*(44). https://doi.org/10.1126/sciadv.abd0288.

2.   United States Environmental Protection Agency. (2019, November). *Advancing Sustainable Materials Management: 2016 and 2017 Tables and Figures*. https://www.epa.gov/sites/default/files/2019-11/documents/2016_and_2017_facts_and_figures_data_tables_0.pdf.

3.   Law, K.L., Starr, N., Siegler, T. R., Jambeck, J. R., Mallos, N. J., & Leonard, G. H. (2020). The United States' Contribution of Plastic Waste to Land and Ocean. *Science Advances, 6*(44). https://doi.org/10.1126/sciadv.abd0288.

4.   Parker, L. (2020, October 30). U.S. Generates More Plastic Trash Than Any Other Nation, Report Finds. *National Geographic*. https://www.nationalgeographic.com/environment/article/us-plastic-pollution.

5.   United States Environmental Protection Agency. (2016). *Hydraulic Fracturing for Oil and Gas: Impacts from the Hydraulic Fracturing Water Cycle on Drinking Water Resources in the United States (Final Report)*. https://cfpub.epa.gov/ncea/hfstudy/recordisplay.cfm?deid=332990.

6. Ibid.

7. Denchak, M. (2019, April 19). Fracking 101. *NRDC*. https://www.nrdc.org/stories/fracking-101#whatis.

8. EHN Editors. (2021, March 1). Fractured: The Body Burden of Living Near Fracking. *Environmental Health News*. https://www.ehn.org/fractured-series-on-fracking-pollution-2650624600/fractured-fracking.

9. Marusic, K. (2021, March 1). Fractured: Harmful Chemicals and Unknowns Haunt Pennsylvanians Surrounded by Fracking. *Environmental Health News*. https://www.ehn.org/fractured-harmful-chemicals-fracking-2650428324/.

10. Center for International Environmental Law (CIEL). (2019, February). *Plastic and Health: The Hidden Costs of a Plastic Planet*. www.ciel.org/plasticandhealth.

11. Ibid.

12. Sullivan, L. (2020, December 22). Big Oil Evaded Regulation and Plastic Pellets. *NPR*. https://www.npr.org/2020/12/22/946716058/big-oil-evaded-regulation-and-plastic-pellets-kept-spilling?sc=18&f=1001.

13. Ibid.

14. Styrene. (2024). In *Wikipedia*. https://en.wikipedia.org/wiki/Styrene.

15. Kim, E. B. (2024, September 30). UC Expert: Growing Oil and Gas Production Means More Styrene, and More Leaks. *The Cincinnati Enquirer*. https://www.cincinnati.com/story/news/2024/09/30/styrene-is-a-byproduct-of-oil-heres-why-thats-important/75415911007/.

16. Ibid.

17. Center for International Environmental Law (CIEL). (2019, February). *Plastic and Health: The Hidden Costs of a Plastic Planet*. www.ciel.org/plasticandhealth.

18. Bilott, R. (2020). *Exposure: Poisoned Water, Corporate Greed, and One Lawyer's Twenty-Year Battle against DuPont*. Atria Books.

19. Steenland, K., & Woskie, S. (2012, November 15). Cohort Mortality Study of Workers Exposed to Perfluorooctanoic Acid. *American Journal of Epidemiology, 176*(10), 909–917. https://doi.org/10.1093/aje/kws171.

20. Ibid.

21. United States Environmental Protection Agency. (2016, May). *Drinking Water Health Advisory for Perfluorooctanoic Acid (PFOA)*. https://www.epa.gov/sites/default/files/2016-05/documents/pfoa_health_advisory_final-plain.pdf.

22. Calafat, A.M., Wong, L.Y., Kuklenyik, Z., Reidy, J.A., & Needham, L.L. (2007, August 29). Polyfluoroalkyl Chemicals in the U.S. Population: Data from the National Health and Nutrition Examination Survey (NHANES) 2003–2004 and Comparisons with NHANES 1999–2000. *Environmental Health Perspectives, 115*(11), 1596–1602. https://doi.org/10.1289/ehp.10598.

23. Lin, A. M, Thompson, J.T., Koelmel, J.P. Liu, Y., Bowden, J.A., and Townsend, T.G. (2024, June 26). Landfill Gas: A Major Pathway for Neutral Per- and Polyfluoroalkyl Substance (PFAS) Release. *Environmental Science & Technology Letters, 11*(7). https://doi.org/10.1021/acs.estlett.4c00364.

24. Karlin, R. (2020, December 3). Cuomo Signs PFAS Ban in Food Packaging. *Times Union.* https://www.timesunion.com/news/article/Cuomo-signs -PFAS-ban-in-food-packaging-15772554.php.

25. Safer States. (n.d.). *Safter States: Bill Tracker, PFAS.* Retrieved March 2021, from https://www.saferstates.org/bill-tracker/?toxic_chemicals=PFAS.

26. Sokolove Law Team. (2020, May 26). As Europe Moves to Ban PFAS, Why Can't the U.S.? *Sokolove Law.* https://www.sokolovelaw.com/blog/ europe-moves-to-ban-pfas/.

27. United States Environmental Protection Agency. (2024). *Key EPA Actions to Address PFAS.* https://www.epa.gov/pfas/key-epa-actions-address-pfas.

28. Ibid.

29. Environmental Health Association of Nova Scotia. (2007, Fall). *Microwave Safe Plastic—Is It Really Safe?* https://www.environmentalhealth.ca/ fall07microwave.html.

30. University of Missouri. (n.d.). *Frederick vom Saal.* Retrieved June 11, 2025, from https://biology.missouri.edu/people/vom-saal.

31. Environmental Health Association of Nova Scotia. (2007, Fall). *Microwave Safe Plastic—Is It Really Safe?* https://www.environmentalhealth.ca/ fall07microwave.html.

32. United States Environmental Protection Agency. (2025, June 3). *Environments and Contaminants—Chemicals in Food.* https://www.epa.gov/ americaschildrenenvironment/environments-and-contaminants -chemicals-food.

33. Mayo Clinic. (2023, March 24). *What Is BPA, and What Are the Concerns about BPA?* https://www.mayoclinic.org/healthy-lifestyle/ nutrition-and-healthy-eating/expert-answers/bpa/faq-20058331.

34. Neltner, T. (2019, August 23). Toxic Chemicals Can Enter Food Through Packaging, so We Made a List *Trellis.* https://www.greenbiz.com/article/ toxic-chemicals-can-enter-food-through-packaging-so-we-made-list.

35. Ibid.

36. Turner, A. (2018). Black Plastics: Linear and Circular Economies, Hazardous Additives and Marine Pollution. *Environment International, 117*, 308–318. https://doi.org/10.1016/j.envint.2018.04.036.

37. Rai, Ayushi. (2018, June 4.) *Down to Earth: Food can have toxic chemicals from recycling e-waste.* https://www.downtoearth.org.in/waste/food-can-have -toxic-chemicals-from-recycling-e-waste-60761.

38. LaMotte, S. (2024, April 1). Flame Retardants Found in Thousands of Consumer Products Linked to Cancer in People for First Time. *CNN.* https://www.cnn.com/2024/04/01/health/flame-retardant-pbdes-cancer -wellness/index.html.

39. Ibid.

40. LaMotte, S. (2024, April 1). Flame Retardants Found in Thousands of Consumer Products Linked to Cancer in People for First Time. *CNN.* https://www.cnn.com/2024/04/01/health/flame-retardant-pbdes-cancer -wellness/index.html.

41. University of Leicester. (2024, May 9). *Colorful Plastics May Lead to More Microplastics: New Study.* https://le.ac.uk/news/2024/may/ microplastic-colours.

42. Laville, S., & Taylor, M. (2017, June 28). A Million Bottles a Minute: World's Plastic Binge "as Dangerous as Climate Change." *The Guardian.* https://www.theguardian.com/environment/2017/jun/28/a-million-a -minute-worlds-plastic-bottle-binge-as-dangerous-as-climate-change.

43. Ibid.

44. Stieger, G. (2017, May 24). Chemicals in Coatings of Coffee-to-Go Cups. *Food Packaging Forum.* https://foodpackagingforum.org/news/ chemicals-in-coatings-of-coffee-to-go-cups.

45. Amaral, J.P. (2020, November 27). Plasticized Childhood: The Impacts of Plastics on Children's Health and the Environment. *Break Free From Plastic.* https://www.breakfreefromplastic.org/2020/11/27/ plasticized-childhood-plastics-impact-children/.

46. Ibid.

47. Elbein, S. (2024, January 11). Plastics Pollution Led to $250 Billion in Disease over One Year. *The Hill.* https://thehill-com.cdn.ampproject .org/c/s/thehill.com/policy/equilibrium-sustainability/4401535-plastics -pollution-disease/amp/.

48. Ibid.

49. Young, J. (2024, November 27). Talks on a Plastics Treaty Have Begun. Here's How to Stop Plastic Waste. *Newsweek.* https://www.newsweek.com/ 2024/12/13/ocean-plastic-pollution-water-microplastics-environment -impact-1991162.html.

50. Kiener, R. (2010). Plastic Pollution. *CQ Researcher, 4*(7). https://doi.org/10 .4135/cqrglobal20100700.

51. EcoWatch. (2018, April 3). *Michigan Lets Nestlé Draw More Groundwater for Bottling.* https://www.ecowatch.com/nestle-water-michigan-2555887499 .html.

52. Kinane, S. (2021, February 3). Will Judge's Ruling Allow Nestlé and Seven Springs to Bottle Nearly a Million Gallons of Water a Day from Florida Springs? *WMNF.* https://www.wmnf.org/nestle-seven-springs-bottle-water -florida-springs/.

53. Stone, M. (2019, August 16). New Plastic Pollution Formed by Fire Looks Like Rocks. *National Geographic.* https://www.nationalgeographic.com/ environment/2019/08/new-plastic-polllution-formed-fire-looks-like-rocks/.

54. Wang, L., Bank, M. S., Rinklebe, J., & Hou, D. (2023). Plastic–Rock Complexes as Hotspots for Microplastic Generation. *Environmental Science & Technology, 57*(17), 7009–7017. https://doi.org/10.1021/acs.est.3c00662.

55. Tangermann, V. (2023, December 19). Scientists Warn That Plastic Rocks Are Appearing Across the World. *The Byte.* https://futurism.com/the-byte/scientists-puzzled-plastic-rocks-world.

56. Oceana. (2021). *Curbing Plastic Pollution at the Source.* Retrieved June 11, 2025, from https://plastics.oceana.org/.

57. Le Page, M. (2018, March 22). The Great Pacific Garbage Patch Is Gobbling Up Ever More Plastic. *NewScientist.* https://www.newscientist.com/article/2164500-the-great-pacific-garbage-patch-is-gobbling-up-ever-more-plastic/.

58. Forbes, M. (Host). (2014, June 26). Garbage Patches: How Gyres Take Our Trash Out to Sea (No. 14) [Audio podcast episode]. In *NOAA Ocean Podcast.* National Ocean Service. https://oceanservice.noaa.gov/podcast/mar18/nop14-ocean-garbage-patches.html.

59. Commonwealth Scientific and Industrial Research Organisation (CSIRO). (2024, April 5). *Ocean Floor a "Reservoir" of Plastic Pollution, Study Finds.* phys.org. https://phys.org/news/2024-04-ocean-floor-reservoir-plastic-pollution.html.

60. Ibid.

61. Oceana. (2021). *Curbing Plastic Pollution at the Source.* Retrieved June 11, 2025, from https://plastics.oceana.org/.

62. Kiener, R. (2010). Plastic Pollution. *CQ Researcher, 4*(7). https://doi.org/10.4135/cqrglobal20100700.

63. Daly, N. (2021, March). For Animals, Plastic Is Turning the Ocean into a Minefield. *National Geographic.* https://www.nationalgeographic.com/magazine/article/plastic-planet-animals-wildlife-impact-waste-pollution.

64. Ibid.

65. Royte, E. (2018, June). We Know Plastic Is Harming Marine Life. What About Us? *National Geographic.* https://www.nationalgeographic.com/magazine/article/plastic-planet-health-pollution-waste-microplastics.

66. Dalhousie University. (2009, March 16). Leatherback Turtles Threatened by Plastic Garbage in Ocean. *Science Daily.* https://www.sciencedaily.com/releases/2009/03/090315224258.htm.

67. Ibid.

68. Ibid.

69. The Center for Biological Diversity. (n.d.). *The Problem With Plastic Bags.* Retrieved June 11, 2025, from, https://www.biologicaldiversity.org/programs/population_and_sustainability/sustainability/plastic_bag_facts.html.

70. McInturf, A., & Savoca, M. (2021, February 9). Hundreds of Fish Species, Including Many That Humans Eat, Are Consuming Plastic. *The Conversation*. https://theconversation.com/hundreds-of-fish -including-many-that-humans-eat-are-consuming-plastic-154634.

71. Hardesty, B. D., Good, T. P., and Wilcox, C. (2015). Novel Methods, New Results, and Science-Based Solutions to Tackle Marine Debris Impacts on Wildlife. *Ocean & Coastal Management, 115*, 4–9. https://doi.org/10.1016/j .ocecoaman.2015.04.004.

72. Baral, S. (2018, December 18). You Can't Just "Clean Up" the Plastic in the Ocean. *Teen Vogue*. https://www.teenvogue.com/story/ you-cant-just-clean-up-the-plastic-in-the-ocean.

73. Elmore, P. B. (2019, October 13). Plastics: A Double-Edged Sword. *Regenration in Action*. https://zerowastezone.blogspot.com/2019/10/plastics -double-edged-sword.html?.

74. Simon, M. (2020, June 11). Plastic Rain Is the New Acid Rain. *Wired*. https://www.wired.com/story/plastic-rain-is-the-new-acid-rain.

75. Ibid.

76. Hundertmark, T., Mayer, M., McNally, C, Simons, T. J., and Witte, C. (2018, December 12). How Plastics Waste Recycling Could Transform the Chemical Industry. *McKinsey & Company*. https://www.mckinsey.com/industries/chemicals/our-insights/ how-plastics-waste-recycling-could-transform-the-chemical-industry#.

77. Vasilyeva, M. (2021, March 19). Is It Snowing Microplastics in Siberia? Russia Scientists Take Samples. *Reuters*. https://www.reuters.com/article/ us-environment-plastic-siberia-snow-idUSKBN2BB19Y.

78. Ruger, A. (2021, December 3). Looking for Microplastics in the Remote Weddell Sea. *Earth.com*. https://www.earth.com/news/ looking-for-microplastics-in-the-remote-weddell-sea/.

79. Wilkinson, F. (2020, November 20). Microplastics Found Near Everest's Peak, Highest Ever Detected in the World. *National Geographic*. https://www .nationalgeographic.com/environment/article/microplastics-found-near -everests-peak-highest-ever-detected-world-perpetual-planet.

80. Godoy, M. (2020, October 19). Study: Plastic Baby Bottles Shed Microplastics When Heated. Should You Be Worried? *Goats and Soda*, NPR. https://www .npr.org/sections/goatsandsoda/2020/10/19/925525183/study-plastic-baby -bottles-shed-microplastics-when-heated-should-you-be-worried.

81. Mercola, J. (2018, April 21). Most Bottled Water Contaminated With Microplastics. *Mercola*. https://articles.mercola.com/sites/articles/ archive/2018/04/21/bottled-water-microplastics-contamination .aspx?v=1617416299.

82. Raza, S. (2024, December 1). Reusing Plastic Water Bottles, To-Go Containers? Scientists Say That's a Bad Idea. *The Washington Post*. https://www.washingtonpost.com/climate-solutions/2024/12/01/ single-use-plastics-reuse-risk/.

83.   Ellen MacArthur Foundation. (2017). *A New Textiles Economy: Redesigning Fashion's Future*. https://content.ellenmacarthurfoundation.org/m/6d507 1bb8a5f05a2/original/A-New-Textiles-Economy-Redesigning-fashions -future.pdf.

84.   De Falco, F., Cocca, M., Avella, M., & Thompson, R. C. (2020). Microfiber Release to Water, Via Laundering, and to Air, via Everyday Use: A Comparison between Polyester Clothing with Differing Textile Parameters. *Environmental Science & Technology, 54*(6). https://doi.org/10.1021/acs .est.9b06892.

85.   Parker, L. (2017, November 30). To Save the Oceans, Should You Give Up Glitter? *National Geographic*. https://www.nationalgeographic.com/science/ article/glitter-plastics-ocean-pollution-environment-spd.

86.   Danopoulos, E., Jenner, L. C., Twiddy, M., & Rotchell, J.M. (2020). Microplastic Contamination of Seafood Intended for Human Consumption: A Systematic Review and Meta-Analysis. *Environmental Health Perspectives, 128*(12). https://doi.org/10.1289/EHP7171.

87.   Milne, M.H., De Frond, H., Rochman, C.M., Mallos, N.J., Leonard, G.H., and Baechler, B.R. (2024, February.) Exposure of U.S. Adults to Microplastics from Commonly-Consumed Proteins. *Environmental Pollution, 343*(15). https://www.sciencedirect.com/science/article/pii/ S0269749123022352.

88.   World Wildlife Foundation. (2019). *No Plastic in Nature: Assessing Plastic Ingestion from Nature to People*. https://wwfint.awsassets.panda.org/ downloads/plastic_ingestion_web_spreads.pdf.

89.   Center for International Environmental Law (CIEL). (2019, February). *Plastic and Health: The Hidden Costs of a Plastic Planet*. www.ciel.org/ plasticandhealth.

90.   Cirino, E. (2019, June 17). You're Likely Inhaling 11 Tiny Bits of Plastic Per Hour. *Vice*. https://www.vice.com/en/article/xwnm74/ youre-likely-inhaling-11-tiny-bits-of-plastic-per-hour.

91.   Wright, S. L., & Kelly, F. J. (2017). Plastic and Human Health: A Micro Issue? *Environmental Science & Technology, 51*(12). https://doi.org/10.1021/ acs.est.7b00423.

92.   Quanquan G., Jiang, J., Huang, Y., Wang, Q., Zhaofeng, L., Ma, X., Yang, X., Li, Y., Wang, W., Cui, W., Tang, J., Wan, H., Xu, Q., Tu, Y., Wu, D., & Xia, Y. (2023). The Landscape of Micron-Scale Particles Including Microplastics in Human-Enclosed Body Fluids. *Journal of Hazardous Materials, 442*. https://doi.org/10.1016/j.jhazmat.2022.130138.

93.   Stone, W. (2024, December 18). Scientists Know Our Bodies Are Full of Microplastics. What Are They Doing to Us? *Shots*, NPR. https://www .npr.org/sections/shots-health-news/2024/12/18/nx-s1-5227172/ microplastics-plastic-nanoparticles-health-pfas.

94.   Adventure Scientists. (2021). *Global Microplastics Initiative*. https://www .adventurescientists.org/microplastics.html.

CHAPTER 5

1. Farmer, J. (1970, April 22). *Earth Day speech.* L'Enfant Plaza, Washington, D.C.

2. Intergovernmental Panel on Climate Change (IPCC). (2018). *Global Warming of 1.5°C.* United Nations. https://www.ipcc.ch/sr15/.

3. Pearce, F. (2019, December 5). As Climate Change Worsens, A Cascade of Tipping Points Looms. *Yale Environment 360.* https://e360.yale.edu/features/as-climate-changes-worsens-a-cascade-of-tipping-points-looms.

4. UN Climate Change. (2019). *UN Climate Change Conference – December 2019.* https://unfccc.int/conference/un-climate-change-conference-december-2019#sessions.

5. U.S. Global Change Research Program. (2009). *Global Climate Change Impacts in the U.S. 2009 Report.* https://web.archive.org/web/20191004023507/https://nca2009.globalchange.gov/global-temperature-and-carbon-dioxide/index.html.

6. Tiseo, I. (2025, January 15). *Average Carbon Dioxide (CO2) Levels in the Atmosphere Worldwide from 1959 to 2024.* Statista. https://www.statista.com/statistics/1091926/atmospheric-concentration-of-co2-historic/.

7. Lindsey, R., & Dahlman, L. (2025, May 29). *Climate Change: Global Temperature.* Climate.gov. https://www.climate.gov/news-features/understanding-climate/climate-change-global-temperature.

8. Borunda, A. (2025, January 10.) 2024 Was the Hottest Year on Record. The Reason Remains a Science Mystery. *NPR.* https://www.npr.org/2025/01/10/nx-s1-5232139/2024-hottest-year-human-history-global-warming.

9. Copernicus. (2025). Surface air temperature for February 2025. Eurpoean Comission. https://climate.copernicus.eu/surface-air-temperature-february-2025.

10. Intergovernmental Panel on Climate Change (IPCC). (2018). *Global Warming of 1.5°C.* United Nations. https://www.ipcc.ch/sr15/.

11. Bamber, J., & Oppenheimer, M. (2019, May 22). Sea Level Rise Could Displace Millions of People Within Two Generations, Study Warns. *CBS News.* https://www.cbsnews.com/news/climate-change-study-sea-level-rise-melting-antartica-ice-displace-millions-two-generations/.

12. Datta, A. (2020, October 20). Here's Why 2020 Is the Worst Year So Far in Terms of Climate Change. *Geospatial World.* https://www.geospatialworld.net/blogs/heres-why-2020-is-the-worst-year-so-far-in-terms-of-climate-change/.

13. McCarthy, J. (2019, April 9). Hunger Looms in Mozambique After Cyclone Idai Wiped Out 1.7 Million Acres of Farmland. *Global Citizen.* https://www.globalcitizen.org/en/content/hunger-crisis-mozambique/.

14. Pulver, D. V., Rice, D., Weise, E., & Padilla, R. (2025, January 10). Earth Passed a Critical Climate Change Threshold in 2024, Scientists Announce. *USA Today.* https://www.usatoday.com/story/news/nation/2025/01/10/2024-hottest-year-annoucement/77543576007/.

15. Wigglesworth, A. (2020, October 3). California Wildfires on the Cusp of Burning 4 Million Acres So Far This Year. *Los Angeles Times*. https://www.latimes.com/california/story/2020-10-03/california-wildfires-burn-nearly-4-million-acres-in-2020.

16. Carrington, D. (2025, March 5). Half of World's CO_2 Emissions Come from 36 Fossil Fuel Firms, Study Shows. *The Guardian*. https://www.theguardian.com/environment/2025/mar/05/half-of-worlds-co2-emissions-come-from-36-fossil-fuel-firms-study-shows.

17. Vinciguerra, T. (interviewer) & Commoner, B. (2007, June 19). At 90, an Environmentalist from the '70s Still Has Hope. *The New York Times*. https://www.nytimes.com/2007/06/19/science/earth/19conv.html.

18. McKibben, B. (2014). *The End of Nature*. Random House Publishing Group.

19. McKibben, B. (2007). *Fight Global Warming Now*. St. Martin's Press.

20. Guggenheim, D. (2006) *An Inconvenient Truth* [Film]. Participant Productions & Lawrence Bender Productions.

21. The Nobel Prize. (n.d.). *Al Gore Facts*. Retrieved June 11, 2025, from https://www.nobelprize.org/prizes/peace/2007/gore/facts/.

22. Kolbert, E. (2006). *Field Notes from a Catastrophe: Man, Nature, and Climate Change*. Bloomsbury.

23. The Cousteau Society. (n.d.). *Our Mission*. Retrieved June 11, 2025, from https://www.cousteau.org/our-mission/.

24. Dicks, D. (1970). *Early Greek Astronomy to Aristotle*. Cornell University Press.

25. Mitchell, M. A. (2025, April 23). Ferdinand Magellan. In *Encyclopedia Britannica*. https://www.britannica.com/biography/Ferdinand-Magellan.

26. Festinger, L. (1962). Cognitive Dissonance. *Scientific American, 207*(4), 93–106. https://doi.org/10.1038/scientificamerican1062-93.

27. Pilat, D., & Krastev, S. (2021). *Why Do We Look for Consistency in Our Beliefs? Cognitive Dissonance Explained*. Retrieved June 11, 2025, from The Decision Lab. https://thedecisionlab.com/biases/cognitive-dissonance/.

28. United Nations Framework Convention on Climate Change. (n.d.). *Key Aspects of the Paris Agreement*. Retrieved June 11, 2025, from https://unfccc.int/process-and-meetings/the-paris-agreement/the-paris-agreement/key-aspects-of-the-paris-agreement.

29. Kopp, O. C. (2025, May 17). Fossil Fuels. In *Encyclopedia Britannica*. Retrieved June 11, 2025, from https://www.britannica.com/science/fossil-fuel.

30. Ibid.

31. Center for International Environmental Law (CIEL). (2019, May). *Plastic & Climate: The Hidden Costs of a Plastic Planet*. www.ciel.org/plasticandclimate.

32. Ibid.

33. USEPA Greenhouse Gas Reporting Program. (2013). *GHGRP Industrial Profiles.* https://www.epa.gov/sites/production/files/2016-11/documents/refineries_2013_112516.pdf.

34. Center for International Environmental Law (CIEL). (2019, May). *Plastic & Climate: The Hidden Costs of a Plastic Planet.* www.ciel.org/plasticandclimate.

35. Moms Clean Air Force. (2019, November 4). *Ethane Cracker Plants: Threatening Our Air, Our Climate, and Our Health.* https://www.momscleanairforce.org/ethane-cracker-plants/.

36. Center for International Environmental Law (CIEL). (2019, May). *Plastic & Climate: The Hidden Costs of a Plastic Planet.* www.ciel.org/plasticandclimate.

37. United States Environmental Protection Agency. (2019, November). *Advancing Sustainable Materials Management: 2016 and 2017 Tables and Figures.* https://www.epa.gov/sites/default/files/2019-11/documents/2016_and_2017_facts_and_figures_data_tables_0.pdf.

38. Center for International Environmental Law (CIEL). (2019, May). *Plastic & Climate: The Hidden Costs of a Plastic Planet.* www.ciel.org/plasticandclimate.

39. YCC Team. (2021, April 28). Air Pollution from Fossil Fuels Caused 8.7 million Premature Deaths in 2018, Study Finds. *Yale Climate Connections.* https://yaleclimateconnections.org/2021/04/air-pollution-from-fossil-fuels-caused-8-7-million-premature-deaths-in-2018-study-finds/.

40. Ibid.

41. Center for International Environmental Law (CIEL). (2019, May). *Plastic & Climate: The Hidden Costs of a Plastic Planet.* www.ciel.org/plasticandclimate.

42. Gordon, K. (2019, September 13). Microplastic Pollution Could Reduce Drinking Water Quality. *BioTechniques.* https://www.biotechniques.com/plant-climate-science/microplastic-pollution-could-reduce-drinking-water-quality/.

43. Laville, S. (2024, November 7). Plastic Pollution Is Changing Entire Earth System, Scientists Find. *The Guardian.* https://www.theguardian.com/environment/2024/nov/07/plastic-pollution-is-changing-entire-earth-system-scientists-find.

44. American Chemistry Council. (n.d.). *Market Access.* Retrieved February 8, 2021, from https://www.americanchemistry.com/better-policy-regulation/trade/market-access.

45. Center for International Environmental Law (CIEL). (n.d.). *Fueling Plastics: Fossils, Plastics, & Petrochemical Feedstocks.* https://www.ciel.org/wp-content/uploads/2017/09/Fueling-Plastics-Fossils-Plastics-Petrochemical-Feedstocks.pdf.

46. Center for International Environmental Law (CIEL). (2019, May). *Plastic & Climate: The Hidden Costs of a Plastic Planet.* www.ciel.org/plasticandclimate.

47. Ibid.

48. Ibid, emphasis added.

49. The Plastics & Climate Project. (2025, May.) *Plastics: Exposing their climate impacts—what we know, what we need to know, & recommendations for research & policy.* https://www.plasticsandclimate.com/publications.

CHAPTER 6

1. Cascio, J. (2009, September 28.) The Next Big Thing: Resilience. *FP.* https://foreignpolicy.com/2009/09/28/the-next-big-thing-resilience/.

2. Container Recycling Institute (CRI). (2021, May 30). *Bottled Water.* https://www.container-recycling.org/index.php/issues/bottled-water.

3. Container Recycling Institute (CRI). (2024, July). *Beverage Container Recycling Fast Facts.* https://www.container-recycling.org/index.php/key-facts.

4. Lee, S. 1990. *The Throwaway Society.* Franklin Watts.

5. Frazier, R. (2019, November 15). The U.S. Natural Gas Boom Is Fueling a Global Plastics Boom. *NPR.* https://www.npr.org/2019/11/15/778665357/the-u-s-natural-gas-boom-is-fueling-a-global-plastics-boom?sc=tw.

6. Crunden, E. A. (2020, February 18). Report Argues Most Plastics, Especially #3-7s, Falsely Labeled as Recyclable. *WasteDive.* https://www.wastedive.com/news/greenpeace-report-recyclables-plastics-circular-economy/572293/.

7. U.S. Securities and Exchange Commission. (2024, September 10). *SEC Charges Keurig with Making Inaccurate Statements Regarding Recyclability of K-Cup Beverage Pod.* https://www.sec.gov/newsroom/press-releases/2024-122.

8. Redd, A. (1995, December 1). Chicago Board of Trade Adds Recyclables. *Waste360.* https://www.waste360.com/mag/waste_chicago_board_trade.

9. The World Bank. (2021, June 5). *The World Bank in China.* https://www.worldbank.org/en/country/china/overview.

10. Maizland, L. (2021, May 19). China's Fight Against Climate Change and Environmental Degradation. *Council on Foreign Relations.* https://www.cfr.org/backgrounder/china-climate-change-policies-environmental-degradation.

11. Yifan, J. (2021, March 25). 14th Five Year Plan: China's Carbon-Centred Environmental Blueprint. *China Dialogue.* https://chinadialogue.net/en/climate/14th-five-year-plan-china-carbon-centred-environmental-blueprint/.

12. National Recycling Coalition. (2017, November 20). *Letter from NRC to World Trade Organization on China and Recycling Issues.* https://nrcrecycles.org/mobius/nrcwp-content/uploads/2018/06/NRC-Letter-to-WTO-November-2017.pdf.

13. Resource Recycling. (2018, February 13). *From Green Fence to Red Alert: A China Timeline.* https://resource-recycling.com/recycling/2018/02/13/green-fence-red-alert-china-timeline/.

14. Butler, N. B. (2018, February 20). In My Opinion: Fix the Broken System. *Plastics Recycling Update.* https://resource-recycling.com/plastics/2018/02/20/opinion-fix-broken-system/.

15. Chiriguayo, D. (2019, April 22). The Single-Use Plastic Pollution Problem. *MPR Marketplace.* https://www.marketplace.org/2019/04/22/single-use-plastic-pollution-problem/.

16. Leahy, S. (2019, July 26). This Common Plastic Packaging Is a Recycling Nightmare. *National Geographic.* https://www.nationalgeographic.com/environment/article/story-of-plastic-common-clamshell-packaging-recycling-nightmare.

17. Ibid.

18. Container Recycling Institute. (2024, July). *Beverage Container Recycling Fast Facts.* https://www.container-recycling.org/images/2024/Fast%20Facts.pdf.

19. Coca-Cola Company. (2024, December 2). *The CocaCola Company Evolves Voluntary Environmental Goals.* https://www.coca-colacompany.com/media-center/the-coca-cola-company-evolves-voluntary-environmental-goals.

20. Parker, L. (2020, September 8). Plastic Food Packaging Now Outpaces Cigarette Butts as Most Abundant Beach Trash. *National Geographic.* https://www.nationalgeographic.com/science/2020/09/plastic-food-packaging-outpaces-cigarette-butts-most-abundant-beach-trash/#close.

21. Keep America Beautiful, Inc. (2021, May). *Keep America Beautiful 2020 National Litter Study.* https://kab.org/litter-study/.

22. SB. 405, Senate Committee on Environmental Quality, 2013–2014 Reg. Sess. (Cal, 2013). http://www.leginfo.ca.gov/pub/13-14/bill/sen/sb_0401-0450/sb_405_bill_20130402_amended_sen_v98.html.

23. World Economic Forum. (2016, January 19). *The New Plastics Economy: Rethinking the Future of Plastics.* https://www.weforum.org/publications/the-new-plastics-economy-rethinking-the-future-of-plastics/.

24. Ibid.

25. Ibid.

26. National Recycling Coalition. (2014, September 16). *NRC Policies.* https://nrcrecycles.org/nrc-policies/.

27. American Chemistry Council. (2021). *What Is Advance Recycling?* https://plastics.americanchemistry.com/what-is-chemical-recycling/?fbclid=IwAR2nffLBzjfgBI7Xij9QOmnTP7Z_1E1YPAwD1oRZ9H9C4qt9yaXG7CG7Cho.

28. Cefic. (2021). *Chemical Recycling Examples.* https://cefic.org/a-solution-provider-for-sustainability/chemical-recycling-making-plastics-circular/chemical-recycling-examples-from-cefic-member-companies/.

29. Tullo, A. H. (2019, October 6). Plastic Has a Problem; Is Chemical Recycling the Solution? *Chemical & Engineering News.* https://cen.acs.org/environment/recycling/Plastic-problem-chemical-recycling-solution/97/i39.

30. GAIA. (2021, February 1). *Letter to Senator Todd Kaminsky*. https://www
    .no-burn.org/.

31. American Chemistry Council. (2021). *What Is Advance Recycling?* https://
    plastics.americanchemistry.com/what-is-chemical-recycling/?fbclid=IwAR2nf
    fLBzjfgBI7Xij9QOmnTP7Z_1E1YPAwD1oRZ9H9C4qt9yaXG7CG7Cho.

32. Bell, L. (2023, October). *Chemical Recycling: A Dangerous Deception*.
    International Pollutants Elimination Network. https://ipen.org/sites/default/
    files/documents/ipen_bp_chemical_recycling_report_11_16_23
    -compressed.pdf.

33. Ibid.

34. GAIA. (2024). *False Solutions to the Plastic Pollution Crisis*. https://www
    .no-burn.org/fact-sheet-false-solutions-to-the-plastic-pollution-crisis/.

35. Ibid.

36. Bell, L. (2023, October). *Chemical Recycling: A Dangerous Deception*.
    International Pollutants Elimination Network.https://ipen.org/sites/
    default/files/documents/ipen_bp_chemical_recycling_report_11_16_23
    -compressed.pdf.

37. Ibid.

38. Rollinson, A. N., & Oladejo, J. (2020). Chemical Recycling: Status,
    Sustainability, and Environmental Impacts. *Global Alliance for Incinerator
    Alternatives*. doi:10.46556/ONLS4535.

39. Tangri, N., & Wilson, M. (2017, March). Waste Gasification &
    Pyrolysis: High Risk, Low Yield Processes for Waste Management.
    Global Alliance for Incinerator Alternatives. https://no-burn.org/
    gasification-pyrolysis-risk-analysis.

40. Quinn, M. (2024, November 26). ExxonMobil, Other
    Chemical Recyclers Announce Latest Investments and
    Partnerships. *WasteDive*. https://www.wastedive.com/news/
    chemical-recycling-2024-updates-alterra-lyondell-basell-purecycle/734018/.

41. World Wildlife Fund. (2022, January 25). *WWF Position: Chemical Recycling
    Implementation Principles*. https://www.worldwildlife.org/publications/
    wwf-position-chemical-recycling-implementation-principles.

42. ReMA Board of Directors. (2022, July 14). *ReMA Announces Position
    on Chemical Recycling*. https://www.isri.org/news-publications/
    news-details/2022/07/29/isri-announces-position-on-chemical-recycling.

43. National Recycling Coalition. (2025). *NRC Chemical Recycling Policy*.
    https://nrcrecycles.org/nrc-home/policy/.

44. United States Envronmental Protection Agency. (2024). *E3 Sustainability
    Tools*. https://www.epa.gov/e3/e3-sustainability-tools#lca.

45. Muralikrishna, I. V., & Manickam, V. (2017). Life Cycle Assessment.
    ScienceDirect. https://www.sciencedirect.com/topics/earth-and-planetary
    -sciences/life-cycle-assessment.

46. Hann, S. (2020, July 28). *Plastics: Can Life Cycle Assessment Rise to the Challenge?* https://www.breakfreefromplastic.org/2020/09/30/can-life-cycle-assessments-rise-to-the-challenge/.

47. Zero Waste International Alliance (ZWIA). (2024, November 2*). Zero Waste Hierarchy of Highest and Best Use 8.0.* https://zwia.org/zwh/.

CHAPTER 7

1. Fuller, R. B. (1969). *Operating Manual for Spaceship Earth.* Vintage.

2. Shiva, D. V. (2014). *Sacred Seed.* The Golden Sufi Center.

3. Space Center Houston. (2019, August 16). *Astronaut Friday: Neil Armstrong.* https://spacecenter.org/astronaut-friday-neil-armstrong/.

4. Rivera, E. (2022, October 23). William Shatner Experienced Profound Grief in Space. It Was the "Overview Effect." *NPR.* https://www.npr.org/2022/10/23/1130482740/william-shatner-jeff-bezos-space-travel-overview-effect?sc=18&f=1001.

5. Jordan, G. (Host), & White, F. (2019, August 30). The Overview Effect (No. 107) [Audio podcast episode]. In *Houston We Have a Podcast.* NASA. https://www.nasa.gov/johnson/HWHAP/the-overview-effect.

6. Stockholm Resilience Centre. (2021, August 25). *Planetary Boundaries.* https://www.stockholmresilience.org/research/planetary-boundaries.html.

7. Ibid.

8. Ibid.

9. Ibid.

10. Ibid.

11. Fuller, R. B. (1969). Spaceship Earth. In *Operating Manual for Spaceship Earth* (pp. 57–65). Vintage.

12. Matthews, D. (2009, June 11). Carbon Emissions Linked to Global Warming in Simple Linear Relationship. *Science Daily.* emphasis added https://www.sciencedaily.com/releases/2009/06/090610154453.htm.

13. Pappas, S. (2017, March 10). Carbon Dioxide Is Warming the Planet. *LiveScience.* https://www.livescience.com/58203-how-carbon-dioxide-is-warming-earth.html.

14. United Nations Framework Convention on Climate Change. (n.d.). *The Paris Agreement.* Retrieved June 11, 2025, from https://unfccc.int/process-and-meetings/the-paris-agreement/the-paris-agreement.

15. SEI, IISD, ODI, E3G, and UNEP. (2020). *The Production Gap Report: 2020 Special Report.* https://productiongap.org/2020report/.

16. Ibid.

17. Intergovernmental Panel on Climate Change. (2021, August 9). *Climate Change Widespread, Rapid, and Intensifying—IPCC.* https://www.ipcc.ch/ 2021/08/09/ar6-wg1-20210809-pr/.

18. Ibid.

19. Sagan, C. (1980, September 28). The Shores of the Cosmic Ocean (Season 1, Episode 1) [TV series episode]. In Andorfer, G., & McCain, R. (Executive Producers), *Cosmos: A Personal Voyage.* KCET.

20. Fischels, J. (2021, June 27). How 165 Words Could Make Mass Environmental Destruction an International Crime. *NPR.* https://www .npr.org/2021/06/27/1010402568/ecocide-environment-destruction -international-crime-criminal-court?sc=18&f=1001.

21. Stop Ecocide Foundation. (2021, June). *Independent Expert Panel for the Legal Definition of Ecocide.* https://static1.squarespace.com/static/ 5ca2608ab914493c64ef1f6d/t/67d4143afd6cf5552db00333 /1741952071334/SE+Foundation+Commentary+and+core+text +2025+texture+print.pdf.

22. Brady, J. (2021, May 26). In a Landmark Case, a Dutch Court Orders Shell to Cut Its Carbon Emissions Faster. *NPR.* https://www.npr.org/2021/05/26/ 1000475878/in-landmark-case-dutch-court-orders-shell-to-cut-its-carbon -emissions-faster.

23. Ibid.

24. Vetter, D. (2021, May 26). "Monumental Victory": Shell Oil Ordered to Limit Emissions in Historic Climate Court Case. *Forbes.* https://www.forbes .com/sites/davidrvetter/2021/05/26/shell-oil-verdict-could-trigger-a-wave -of-climate-litigation-against-big-polluters/?sh=196f176d1a79.

25. Ibid.

26. Fawcett-Atkinson, M. (2021, May 13). Canada Officially Tosses Plastic in the "Toxic" Bin. *Canada's National Observer.* https://www.nationalobserver .com/2021/05/13/news/canada-officially-tosses-plastic-toxic-bin.

27. Ibid.

28. Kusnetz, N. (2024, April 4). Should Big Oil Be Tried for Homicide? *Inside Climate News.* https://insideclimatenews.org/news/04042024/ fossil-fuel-companies-homicide-charge/.

29. Fawcett-Atkinson, M. (2021, May 13). Canada Officially Tosses Plastic in the "Toxic" Bin. *Canada's National Observer.* https://www.nationalobserver .com/2021/05/13/news/canada-officially-tosses-plastic-toxic-bin.

30. Intergovernmental Panel on Climate Change. (2021, August 9). *Climate Change Widespread, Rapid, and Intensifying.* https://www.ipcc.ch/2021/ 08/09/ar6-wg1-20210809-pr/.

31. McMurtry, A. (2021, December 5). *Greenhouse Gas Emissions Are Shrinking the Stratosphere.* Anadolu Agency. https://www.aa.com.tr/en/environment/ greenhouse-gas-emissions-are-shrinking-the-stratosphere-study/2239327.

32. World Meteorological Organization. (2020, March 10). Flagship UN Study Shows Accelerating Climate Change on Land, Sea and in the Atmosphere. *UN News.* https://news.un.org/en/story/2020/03/1059061.

33. Ibid.

34. Thunberg, G. (2019). *No One Is Too Small to Make a Difference.* Penguin Books.

CHAPTER 8

1. Fuller, R. B. (1977). In Peter, L. J. *The Peter Plan: A Proposal for Survival.* Bantam.

2. Contrarian. (2022). In *Merriam-Webster Dictionary.* https://www.merriam -webster.com/dictionary/contrarian.

3. Sample, S. B. (2002). *Contrarian's Guide to Leadership.* Jossey-Bass.

4. Ibid.

5. Ibid.

6. Priestley, D. (2015, November 30). Binary Thinking vs Directional Thinking. *Key Person of Influence* (Blog), Dent. http://www .keypersonofinfluence.com/binary-thinking-vs-directional-thinking/.

7. Ibid.

8. Iannacci, N. (2016, April 12). Henry Clay, the Great Compromiser. *National Constitution Center.* https://constitutioncenter.org/blog/ henry-clay-the-great-compromiser.

9. History of the United States debt ceiling. (2023, June 1). In *Wikipedia.* https://en.wikipedia.org/wiki/History_of_the_United_States_debt_ceiling.

10. Baker, P. (2023, June 1). The Calm Man in the Capital: Biden Lets Others Spike the Ball but Notches a Win. *The New York Times.* https://www .nytimes.com/2023/06/01/us/politics/biden-mccarthy-debt-ceiling-deal -who-won.html?searchResultPosition=1.

11. Sánchez, G.J., Jarenwattananon, P., & Chang, A. (2023, June 1). Sen. Jeff Merkley from Oregon Opposes the Debt Ceiling Bill Heading to the Senate. *NPR.* https://www.wwno.org/npr-news/npr-news/2023-06-01/sen-jeff -merkley-from-oregon-opposes-the-debt-ceiling-bill-heading-to-the-senate.

12. Ibid.

13. Rappeport, A. (2023, May 27). Yellen's Debt Limit Warnings Went Unheeded, Leaving Her to Face Fallout. *The New York Times.* https://www .nytimes.com/2023/05/27/us/politics/yellen-debt-limit.html.

14. Sierra Club. (2021, August 27). *Six Years and $100 Million Later, Ohio Petrochemical Plant Delayed Yet Again.* https://www.sierraclub.org/press -releases/2021/08/six-years-and-100-million-later-ohio-petrochemical-plant -delayed-yet-again.

15. Smith, A. (1776). *The Wealth of Nations*. W. Strahan and T. Cadell.

16. Gilchrist, T. (2006, December 15). Interview: Mel Gibson. *IGN*. https://www.ign.com/articles/2006/12/15/interview-mel-gibson.

17. McLaren, B. D. (2009). *Everything Must Change*. Thomas Nelson.

18. Organization for Economic Co-operation and Development. (2022, February 22). *Plastic Pollution Is Growing Relentlessly as Waste Management and Recycling Fall Short, Says OECD*. https://www.oecd.org/environment/plastic-pollution-is-growing-relentlessly-as-waste-management-and-recycling-fall-short.htm.

19. Ibid.

20. UN Environment Programme. (2025). *Our Planet Is Choking on Plastic*. https://www.unep.org/interactives/beat-plastic-pollution/.

21. Smythe, K. R. (2020). *Whole Earth Living: Reconnecting Earth, History, Body, and Mind*. London, England: Dixi Books Publishing.

22. Ibid.

23. Bambridge-Sutton, A. (2023, November 14). Beverage Giants in Hot Water Over Plastic Bottle Recycling Claims. *Food Navigator*. https://www.foodnavigator.com/Article/2023/11/14/nestle-danone-and-the-coca-cola-company-in-hot-water-over-plastic-bottle-recycling-claims/.

24. Daly, H. E. (1991). *Steady State Economics*. Island Press.

25. United Nations Department of Economic and Social Affairs. (n.d.). *The 17 Goals*. Retrieved June 11, 2025, from https://sdgs.un.org/goals.

CHAPTER 9

1. Roosevelt, E. (1960). *You Learn by Living: Eleven Keys for a More Fulfilling Life*. Westminster John Knox Press.

2. Ammer, C. (2013). Waste Not, Want Not. In *American Heritage Dictionary of Idioms* (2nd ed.). Houghton Mifflin Harcourt.

3. Fishman, K., & McKee, J. (2015, January 15). Scrap for Victory! *Library of Congress Blogs*. https://blogs.loc.gov/now-see-hear/2015/01/scrap-for-victory/.

4. Franklin D. Roosevelt Presidential Library and Museum. (n.d.). Use It Up, Wear It Out, Make It, Do, or Do Without. https://www.fdrlibrary.org/use-it-up-wear-it-out-make-it-do-or-do-without.

5. Freinkel, S. (2011). *Plastic: A Toxic Love Story*. Houghton Mifflin Harcourt.

6. Ibid.

7. Intergovernmental Panel on Climate Change (IPCC). (2023, March 20). *Synthesis Report of the Sixth Assessment Report*. https://www.ipcc.ch/report/ar6/syr/resources/spm-headline-statements/.

8. Ibid.

9. Tiseo, I. (2021, January 27). *Global Market Value of Plastic 2018–2027*. Statista. https://www.statista.com/statistics/1060583/global-market-value -of-plastic/#statisticContainer.

10. Ibid.

11. Ritchie, H. (2022, January 11). How the World Eliminated Lead from Gasoline. *Our World in Data*. https://ourworldindata.org/leaded-gasoline -phase-out.

12. Radcliffe, S. (2021, August 29). Banning CFCs Helped Us Avoid an Even Worse Climate Catastrophe. *Healthline*. https://www.healthline.com/health -news/banning-cfcs-helped-us-avoid-an-even-worse-climate-catastrophe.

13. Deutsche Welle. (2016, October 15). *Nearly 200 Nations Reach HFC Phase-Out Deal*. https://www.dw.com/en/nearly-200-nations-reach-agreement-to -phase-out-hfc-greenhouse-gases/a-36049841.

14. Ritchie, H. (2022, January 11). How the World Eliminated Lead from Gasoline. *Our World in Data*. https://ourworldindata.org/leaded-gasoline -phase-out.

15. U.S. Department of Energy. (n.d.). *Alternative Fuels Data Center— Renewable Gasoline*. Alternative Fuels Data Center. Retrieved June 16, 2025, from https://afdc.energy.gov/fuels/emerging_hydrocarbon.html.

16. Nilsen, E. (2023, March 14). The Willow Project Has Been Approved. Here's What to Know About the Controversial Oil-Drilling Venture. *CNN*. https://www.cnn.com/2023/03/14/politics/willow-project-oil-alaska -explained-climate/index.html.

17. Gardiner, B. (2019, December 19). The Plastics Pipeline: A Surge of New Production Is on the Way. *Yale Environment 360*. https://e360 .yale.edu/features/the-plastics-pipeline-a-surge-of-new-production -is-on-the-way.

18. UN Environment Programme. (2025). *Intergovernmental Negotiating Committee on Plastic Pollution*. https://www.unep.org/inc-plastic-pollution.

19. Linko, E. (2023). *International Panel on Climate Change Working Group III*. Xavier University.

20. Center for International Environmental Law (CIEL). (2019, May). *Plastic & Climate: The Hidden Costs of a Plastic Planet*. www.ciel.org/plasticandclimate.

21. Intergovernmental Panel on Climate Change (IPCC). (2022). *Climate Change 2022: Mitigation of Climate Change*. https://www.ipcc.ch/report/ sixth-assessment-report-working-group-3/.

22. Center for International Environmental Law. (2017). *Fueling Plastics: How Fracked Gas, Cheap Oil, and Unburnable Coal Are Driving the Plastics Boom*. https://www.ciel.org/wp-content/uploads/2017/09/Fueling-Plastics-How -Fracked-Gas-Cheap-Oil-and-Unburnable-Coal-are -Driving-the-Plastics-Boom.pdf.

23.   Ibid.

24.   Secretariat of the Basel Convention. (2011). *Basel Convention on the Control and Transboundary Movement of Hazardous Wastes and Their Disposal.* http://www.basel.int/TheConvention/Overview/tabid/1271/Default.aspx.

25.   Ibid.

26.   Intergovernmental Panel on Climate Change (IPCC). (2023, March 20). *Urgent Climate Action Can Secure a Liveable Future for All.* https://www.ipcc.ch/2023/03/20/press-release-ar6-synthesis-report/.

27.   Brown, A. (2025, January 30). *Unlocking Economic Opportunities with Non-Plastic Substitutes: A Path to Sustainable Growth.* LinkedIn. https://www.linkedin.com/pulse/unlocking-economic-opportunities-non-plastic-path-growth-brown-mxpie.

28.   Organisation for Economic Co-operation and Development. (2022, February 22). *Plastic Pollution Is Growing Relentlessly as Waste Management and Recycling Fall Short, Says OECD.* https://www.oecd.org/en/about/news/press-releases/2022/02/plastic-pollution-is-growing-relentlessly-as-waste-management-and-recycling-fall-short.html.

29.   Rocha, M. (2019, October 9). *Listen to Willy Wonka: A Lesson in Consumerism.* Medium. https://malu-rocha.medium.com/listen-to-willy-wonka-a-lesson-in-consumerism-6d52a6d59b9d.

30.   Upstream. (2023). *Upstream Research.* https://upstreamsolutions.org/research.

31.   Rabson, M. (2022, June 20). Say Goodbye to Some Single-Use Plastics as Federal Ban Is Phased In. *Canada's National Observer.* https://www.nationalobserver.com/2022/06/20/news/single-use-plastics-federal-ban.

32.   César E. Chávez National Monument. (2025, March 20). *Workers United: The Delano Grape Strike and Boycott.* National Park Service. https://www.nps.gov/articles/000/workers-united-the-delano-grape-strike-and-boycott.htm.

33.   World Wildlife Fund (WWF). (2022, December 7). *Transparent 2022: Annual ReSource: Plastic Progress Report.* https://www.worldwildlife.org/publications/transparent-2022-annual-resource-plastic-progress-report.

34.   National Recycling Coalition. (2017, June 9). *NRC Policy Document #43—Reuse.* https://docs.google.com/document/d/1HwbzjdigxsyBIyNxzuU2LRMkyaMtyRqcJ66v8QAvD_Q/edit.

35.   Upstream. (2024). *Switch to Reuse.* https://upstreamsolutions.org/business-home.

36.   U.S. Plastics Pact. (2023). *The U.S. Plastics Pact Roadmap to 2025.* https://usplasticspact.org/roadmap/.

37.   U.S. Plastics Pact. (2022, October 7). *Reuse & Refill Sustainable Packaging Innovation Award Finalists.* https://usplasticspact.org/reuse-refill-2022-award-finalists/.

38. Closed Loop Partners and U.S. Plastics Pact. (2025, March 11). *Closed Loop Partners and U.S. Plastics Pact Identify Top 5 Consumer Product Categories Poised for Near-Term Reuse Success in U.S. Retail.* https://www.closedlooppartners.com/top-5-product-categories-for-reuse/.

39. DeGroff, K. (2022, Fall). Disposing Disposables. *National Parks Conservation Association.* https://www.npca.org/articles/3270-ditching-disposables.

40. Ibid.

41. Upsteam. (2022). *The Reusies.* https://www.thereusies.org/home-2.

42. Fixit Clinic. (2023). *About.* https://fixitclinic.blogspot.com/p/bring-your-broken-non-functioning.html.

43. Ogle, A. (2022, June 24). The 7 Best Reusable Utensils of 2023. *Treehugger.* https://www.treehugger.com/best-reusable-utensils-5087056.

44. Alexander, S. (2022, September 2). First Person Perspective: A "Pitiful" Attack on Plastics Recycling. *Resource Recycling.* https://resource-recycling.com/recycling/2022/09/02/first-person-perspective-a-pitiful-attack-on-plastics-recycling/.

45. The Maritime Executive. (2021, March 14). *Report: U.S. Plastic Waste Exports May Violate Basel Convention.* https://maritime-executive.com/article/report-u-s-plastic-waste-exports-may-violate-basel-convention.

46. Container Recycling Institute (CRI). (2024). *Redemption Rates and Other Features of 10 U.S. State Deposit Programs.* https://www.bottlebill.org/Allstates/10%20states%20Redemption%20Rates%20102924.pdf.

47. U.S. Plastics Pact. (2023). *Roadmap to 2025.* https://usplasticspact.org/roadmap/.

48. Ocean Titans. (2023). *World Oceans Day.* https://www.theoceantitans.com/.

49. Ocean Plastics Recovery Project. (2023). *Our Story.* https://oceanplasticsrecovery.com/story.

50. Oceanworks. (2023). *Sustainable Material with Reliability at Scale.* https://oceanworks.co/pages/materials.

51. Baker, R. (2022, June 30). Recycling Isn't the "Panacea" That Saves Oceans from Plastic. *Canada's National Observer.* https://www.nationalobserver.com/2022/06/30/news/recycling-isnt-panacea-saves-oceans-plastic-un-conference.

52. Oxford University Press. (n.d.). Skeuomorph, n. In *Oxford English Dictionary.* https://doi.org/10.1093/OED/1114685993.

53. Freinkel, S. (2011). *Plastic: A Toxic Love Story.* Houghton Mifflin Harcourt.

54. Ghaddar, A., & Bousso, R. (2018, October 5). Rising Use of Plastics to Drive Oil Demand to 2050: IEA. *Reuters.* https://www.reuters.com/article/business/environment/rising-use-of-plastics-to-drive-oil-demand-to-2050-iea-idUSKCN1ME2QC/.

55. U.S. Plastics Pact. (2024). *Problematic and Unnecessary Materials Report.* https://usplasticspact.org/wp-content/uploads/dlm_uploads/2022/01/U.S. -Plastics-Pact-Problematic-Unnecessary-Materials-Report-1.25.2022.pdf.

56. Milet, K. (2022, October 3). Can Cellulose Replace Plastic in Packaging? *Premium Beauty News.* https://www.premiumbeautynews.com/en/ can-cellulose-replace-plastic-in,20908.

57. Ibid.

58. Grenoble INP Foundation. (2021). *Cellulose Valley Chair.* https://fondation -grenoble-inp.fr/en/nos-actions/cellulose-valley/.

59. Clancy, H. (2024, July 10). Lego Sets Stricter Emissions Reductions Requirements for Suppliers. *Trellis.* https://trellis.net/article/ lego-sets-stricter-emissions-reductions-requirements-suppliers/.

60. Plastics Industry Association. (n.d.). *Bioplastics.* Retrieved June 16, 2025, from https://www.plasticsindustry.org/supply-chain/recycling-sustainability/ bioplastics.

61. Youth BioTech Journal (n.d.). *Bioplastics – What Are They Really?* Retrieved June 16, 2025, from https://www.ybtjournal.com/post/ bioplastics-what-are-they-really.

62. Siegel, R. (2019, May 8). The Rise of Plant-Based Plastic Packaging. *Trellis.* https://trellis.net/article/rise-plant-based-plastic-packaging/.

63. Karidis, A. (2021, April 1). Why Mars Wrigley Is Turning to Compostable Packaging and Future Ambitions. *Waste360.* https://www.waste360.com/ sustainability/why-mars-wrigley-turning-compostable-packaging-and-future -ambitions?NL=WST-03&Issue=WST-03_202%E2%80%A6.

64. Ibid.

65. Petsko, E. (2020, July 21). Recycling Myth of the Month: Plant-Based Bioplastics Are Not as 'Green' as Some Think. *Oceana.* https://oceana.org/ blog/recycling-myth-month-plant-based-bioplastics-are-not-green-some -think.

66. BPI. (2023). *Homepage.* https://bpiworld.org/.

67. Leffer, L. (2021, August 14). Your Compostable Cups and Containers Aren't Reversing the Plastic Problem. *Popular Science.* https://www.popsci.com/ environment/truth-about-compostable-cups/.

68. Eunomia Research and Consulting. (2024). *Bioplastics Are Trash: The Unforeseen Environmental Consequences of PLA from Production to Disposal.* https://eunomia.eco/reports/bioplastics-are-trash-the-unforeseen -environmental-consequences-of-pla-from-production-to-disposal/.

69. Ibid.

70. Rolsky, C. and Kelkar, V. (2021, June 3). Degradation of Polyvinyl Alcohol in US Wastewater Treatment Plants and Subsequent Nationwide Emission Estimate. *International Journal Environmental Research and Public Health,* *18*(11): 6027. https://pmc.ncbi.nlm.nih.gov/articles/PMC8199957/.

71. Plastic Pollution Coalition. (2022, November 18). *PVA Plastic: What You Need to Know.* https://www.plasticpollutioncoalition.org/blog/2022/11/18/pva-plastic-what-you-need-to-know.

72. Ibid.

73. Project Drawdown. (2023). *Bioplastics.* https://drawdown.org/solutions/bioplastics.

74. Seth Borenstein, M. W. (2023, July 3). Climate Change Effects Getting Worse. *The Cincinnati Enquirer,* pp. A10.

75. Walling, S. B. (2023, July 7). Earth Remains Under Record Heat Wave. *The Cincinnati Enquirer,* pp. A08.

76. Ibid.

77. Seth Borenstein, M. W. (2023, July 3). Climate Change Effects Getting Worse. *The Cincinnati Enquirer,* pp. A10.

78. U.S. Department of Health and Human Services. (n.d.). *Public Health & Safety.* Retrieved June 16, 2025, from https://www.hhs.gov/programs/public-health-safety/index.html.

79. Occupational Safety and Health Administration Education Center (OSHA). (n.d.). *How These Safety & Health Misconceptions Are Slowing Your Growth.* Retrieved June 16, 2025, from https://www.oshaeducationcenter.com/articles/safety-and-health-misconceptions/.

80. U.S. Environmental Protection Agency (EPA). (n.d.). *Regulations.* Retrieved June 16, 2025, from https://www.epa.gov/laws-regulations/regulations.

81. Halt the Harm Network. (n.d.). *Fracking Chemicals Near You: OpenFF Consultations.* Retrieved June 16, 2025, from https://halttheharm.net/fracking-chemicals-near-you/.

82. Plastic Free Future. (n.d.). *Sustainability Consulting Services.* Retrieved June 16, 2025, from https://plastic-free-future.org/programs-and-services.

83. Ellen MacArthur Foundation. (2023, July 13). *What Is a Circular Economy?* https://ellenmacarthurfoundation.org/topics/circular-economy-introduction/overview.

84. Ibid.

85. Riddle, M. (2022, October 31). Reality Check: Circular Plastic Is a Myth. *Triple Pundit.* https://www.triplepundit.com/story/2022/circular-plastic-myth/758721.

86. Peryman, M. (2023, May 23). Why Plastics Can Never Be "Circular": A Māori Perspective on the Global Plastics Treaty. *Greenpeace.* https://www.greenpeace.org/aotearoa/story/why-plastics-can-never-be-circular/.

87. United States Environmental Protection Agency. (2023, July 13). *What Is a Circular Economy?* https://www.epa.gov/circulareconomy/what-circular-economy.

88. Vidal, F., van der Marel, E. R., Kerr, R. W. F., McElroy, C., Schroeder, N., Mitchel, C., Rosetto, G., Chen, T. T. D., Bailey, R. M., Hepburn, C., Redgwell, C., & Williams, C. K. (2024). Designing a circular carbon and plastics economy for a sustainable future." *Nature*, 626, 45–57. https://doi.org/10.1038/s41586-023-06939-z.

89. Intergovernmental Panel on Climate Change (IPCC). (2024). *AR6 Synthesis Report: Climate Change 2023*. https://www.ipcc.ch/report/sixth-assessment-report-cycle/.

CHAPTER 10

1. Weldon, F. (1991). Subject to Diary. In *Moon Over Minneapolis*. Harper Collins.

2. Intergovernmental Panel on Climate Change (IPCC). (2024). AR6 Synthesis Report: Climate Change 2023. https://www.ipcc.ch/report/sixth-assessment-report-cycle/.

3. Gandhi, M. (1940, September 1). *Mohan-Mala (A Gandhian Rosary)*. https://www.mkgandhi.org/mohanmala/june.php.

4. City of Austin. (2010, April). *What's YOUR One Green Step?* https://archive.org/details/auscitx-What_s_YOUR_One_Green_Step.

5. Prengaman, P. (2023, September 15). Citing Sustainability, Starbucks Wants to Overhaul Its Iconic Cup. Will Customers Go Along? *Associated Press*. https://apnews.com/article/starbucks-cup-disposable-sustainable-climate-5a3a5d7d4a66725f9747cbab23cb586d.

6. Young, R., Healy, G., & Hagan, A. (2024, January 4). Glitter's Microplastic Problem: The Environmental Case for Breaking Up with Glitter in Makeup. *WBUR-NPR Here & Now*. https://www.wbur.org/hereandnow/2024/01/04/glitter-makeup-microplastics.

7. Wong, V. (n.d.). *Hi, I'm Von Wong, and I Make Your Impact Unforgettable*. Unforgettable. Retrieved June 16, 2025, from https://unforgettablelabs.com/.

8. Upstream. (2024, April 25). *Big Brands Can—and Must—Be Changemakers*. https://upstreamsolutions.org/blog/big-brands-can-and-must-be-changemakers.

9. DeLuca, A. (2024, January). *Delta Chief Sustainability Officer*. LinkedIn. https://www.linkedin.com/feed/update/urn:li:activity:7137820989505110016/.

10. Earth911. (n.d.). *Earth911*. Retrieved June 16, 2025, from https://earth911.com/.

11. Aslanian, S. (2015, February 15). Fix-It Clinics Bring the Broken to Life—and Cut Waste. *MPRNews*. https://www.mprnews.org/story/2015/02/12/fix-it-clincs.

12. Local Tools. (n.d.). *The Easy Way to Manage a Lending Library.* Retrieved June 16, 2025, from https://localtools.org/.

13. Hill, N. (2007). *Think and Grow Rich: The Landmark Bestseller—Now Revised and Updated for the 21st Century.* Tarcher.

14. Earthday.org. (n.d.). *Pick an action.* Retrieved June 16, 2025, from https://action.earthday.org/.

15. Plastic Free Foundation. (n.d.). *Plastic Free July.* Retrieved June 16, 2025, from https://www.plasticfreejuly.org/.

16. U.S. Conference of Mayors. (2021). *2021 Adopted Resolutions: Environment.* https://legacy.usmayors.org/resolutions/89th_Conference/proposed-review-list-full-print-committee.asp?committee=Environment.

17. Charron, A. (2023, August 2). *End Plastics: 60×40: A Global Wave to Halt Plastic Production and Save Our Planet.* EarthDay.org. https://www.earthday.org/60x40-a-global-wave-to-halt-plastic-production-and-save-our-planet/.

18. American Medical Association. (2022, June 13). *AMA Adopts New Policy Declaring Climate Change a Public Health Crisis.* https://www.ama-assn.org/press-center/press-releases/ama-adopts-new-policy-declaring-climate-change-public-health-crisis.

19. LA Sanitation and Environment. (2024, March). *Comprehensive Plastics Reduction Program.* https://sanitation.lacity.gov/san/faces/home/portal/s-lsh-es/s-lsh-es-ceqap/s-lsh-es-ceqap-cprp?_adf.ctrl-state=15tugazwgp_5&_afrLoop=15065592204398351#!.

20. Halt the Harm Network. (n.d.). *Connect With Others Fighting Oil & Gas Industry Pollution.* Retrieved June 16, 2025, from https://halttheharm.net/.

21. Smith, A. (2024, July 1). PepsiCo, Coca-Cola Sued by Baltimore for Cleanup. *Resource Recycling.* https://resource-recycling.com/recycling/2024/07/01/pepsico-coca-cola-sued-by-baltimore-for-cleanup-costs/; Bisset, B. V. (2024, June 22). Young Climate Activists Just Won a "Historic" Settlement. *Washington Post.* https://www.msn.com/en-us/news/us/young-climate-activists-just-won-a-historic-settlement/ar-BB1oER1a; Brugger, J. (2024, June 5). Lawsuits Targeting Plastic Pollution. *Insude Climate News.* https://insideclimatenews.org/news/05062024/lawsuits-targeting-plastic-pollution-pile-up-as-frustrated-citizens-and-states-seek-accountability; Associated Press. (2023, August 14). Judge Sides With Young Activists in First-of-Its-Kind Climate Change Trial in Montana. *NPR.* https://www.npr.org/2023/08/14/1193780700/montana-climate-change-trial-ruling; Associated Press. (2022, October 18). NJ Sues Oil, Gas Firms, Trade Group Over Climate Change. *AP News.* https://apnews.com/article/business-lawsuits-new-jersey-city-exxon-mobil-corp-ebd2e5381054bc4d5da50e1478eab35f.

22. Greenfield, P. (2023, January 18). Revealed: More Than 90% of Rainforest Carbon Offsets by Biggest Certifier Are Worthless, Analysis Shows. *The Guardian.* https://www.theguardian.com/environment/2023/jan/18/revealed-forest-carbon-offsets-biggest-provider-worthless-verra-aoe.

23. Blue Green Alliance. (n.d.). *Home*. Retrieved June 16, 2025, from https://www.bluegreenalliance.org/.

24. A New Earth Project. (n.d.). *Shop—New Earth Approved*. Retrieved June 16, 2025, from https://anewearthproject.com/collections/new-earth-approved.

25. Future Coalition. (n.d.). *Action Center*. Retrieved June 16, 2025, from https://futurecoalition.org/actions/.

26. Gen-Z for Change. (n.d.). *Home*. Retrieved June 16, 2025, from https://www.genzforchange.org/.

27. Zero Hour (n.d.). *Support Zero Hour and Amplify the Voices of Youth Organizing for Climate Action*. Retrieved June 16, 2025, from https://thisiszerohour.org/.

28. Black Girl Environmentalist [@blackgirlenvironmentalist]. (n.d.). *Home* [Instagram profile]. Retrieved June 16, 2025, from https://www.instagram.com/blackgirlenvironmentalist/.

29. Our Children's Trust. (n.d.). *Home*. Retrieved June 16, 2025, from https://www.ourchildrenstrust.org/.

30. Third Act. (n.d.). *Home*. Retrieved June 16, 2025, from https://thirdact.org/.

31. Science Moms. (n.d.). *Home*. Retrieved June 16, 2025, from https://sciencemoms.com/climate-scientists/.

32. Change the Chamber. (n.d.). *From Doom and Gloom to a Sustainable Future*. Retrieved June 16, 2025, from https://www.changethechamber.org./

33. #BreakFreeFromPlastic. (n.d.). *The Global Movement Envisioning a Future Free From Plastic Pollution*. Retrieved June 16, 2025, from https://www.breakfreefromplastic.org.

34. 350.org. (n.d.). *We Fight for a World beyond Fossil Fuels*. Retrieved June 16, 2025, from https://350.org/us-homepage/.

35. United States Environmental Protection Agency. (2025, February 14). *Draft National Strategy to Prevent Plastic Pollution*. https://www.epa.gov/circulareconomy/draft-national-strategy-prevent-plastic-pollution.

36. Karidis, A. (2022, March 25). The UK Follows EU With Plastic Tax. *Waste360*. https://www.waste360.com/plastics/the-uk-follows-eu-with-plastic-tax.

37. National Stewardship Action Council (NSAC). (n.d.). *Home*. Retrieved June 16, 2025, from https://www.nsaction.us/.

38. Container Recycling Institute. (n.d.). *Home*. Retrieved June 16, 2025, from https://www.container-recycling.org/index.php.

39. Beyond Plastics. (n.d.). *Synthetic Turf Is HAZARDOUS*. Retrieved June 16, 2025, from https://www.beyondplastics.org/fact-sheets/synthetic-turf; Voos, J. E. (2019, August 26). *Artificial Turf Versus Natural Grass*. University Hospitals. https://www.uhhospitals.org/for-clinicians/articles-and-news/articles/2019/08/artificial-turf-versus-natural-grass.

40. As You Sow. (2024, June 11). *2024 Plastic Promises Scorecard.* https://www
    .asyousow.org/reports/2024-plastic-promises-scorecard.

41. Intergovernmental Panel on Climate Change (IPCC). (n.d.). *IPCC AR6
    Synthesis Report: Climate Change 2023.* Retrieved June 16, 2025, from
    https://www.ipcc.ch/report/ar6/syr/.

42. UN Environment Programme. (n.d.). *Plastic Pollution.* Retrieved June 16,
    2025, from https://www.unep.org/plastic-pollution.

43. UN Environment Programme. (2021, October 21). *From Pollution to
    Solution: A Global Assessment of Marine Litter and Plastic Pollution.* https://
    www.unep.org/resources/pollution-solution-global-assessment-marine-litter
    -and-plastic-pollution.

44. UN Environment Programme. (2022, March 2). *Historic Day in the
    Campaign to Beat Plastic Pollution: Nations Commit to Develop a Legally
    Binding Agreement.* https://www.unep.org/news-and-stories/press-release/
    historic-day-campaign-beat-plastic-pollution-nations-commit-develop.

45. Basel Convention on the Control of Transboundary Movements of
    Hazardous Wastes and their Disposal. (n.d.). *Overview.* Retrieved June 16,
    2025, from https://www.basel.int/Implementation/Plasticwaste/Overview/
    tabid/8347/Default.aspx.

46. U.S. Department of State. (n.d.). *Basel Convention on Hazardous Wastes.*
    https://www.state.gov/key-topics-office-of-environmental-quality
    -and-transboundary-issues/basel-convention-on-hazardous-wastes/.

47. The Fossil Fuel Non-Proliferation Treaty Initiative. (n.d.). *Join the Call
    for a Fossil Fuel Treaty to Manage a Global Transition to Safe, Renewable
    & Affordable Energy for All.* Retrieved June 16, 2025, from https://
    fossilfueltreaty.org/.

48. U.S. Plastics Pact. (n.d.). *U.S. Plastics Pact Problematic and Unnecessary
    Materials Report.* Retrieved June 16, 2025, from https://usplasticspact.org/
    problematic-materials/.

49. UN Environment Programme. (n.d.). *About.* Retrieved June 16, 2025, from
    https://leap.unep.org/en/content/basic-page/plastics-pollution-toolkit-about.

50. European Enrironmental Bureau. (2024, June 13). *EU Has "Legal
    Duty" to Ban PVC, NGOs Tell European Commission.* https://eeb.org/
    eu-has-legal-duty-to-ban-pvc-ngos-tell-european-commission/.

51. *Break Free From Plastic Pollution Act of 2023.* S.3127, 118th Cong. (2023–
    2024). https://www.congress.gov/bill/118th-congress/senate-bill/3127.

52. *Green Climate Fund Authorization Act of 2023.* H.R.3961, 118th
    Cong. (2023–2024). https://www.congress.gov/bill/118th-congress/
    house-bill/3961.

53. *Earth Act to Stop Climate Pollution by 2030.* H.R.598, 118th Cong. (2023–
    2024). https://www.congress.gov/bill/118th-congress/house-bill/598.

54. *Climate Displaced Persons Act.* S.3340, 118th Cong. (2023–2024). https://
    www.congress.gov/bill/118th-congress/senate-bill/3340.

55. *Reducing Waste in National Parks Act.* H.R.4561, 118th Cong. (2023–2024). https://www.congress.gov/bill/118th-congress/house-bill/4561.

56. Container Recycling Institute. (n.d.). *Home.* Retrieved June 16, 2025, from https://www.bottlebill.org/; Can Manufacturers Institute. (n.d.). *Recycling Refunds.* Retrieved June 16, 2025, from https://www.cancentral.com/recyclingrefunds/; National Recycling Coalition. (n.d.). *NRC Container Deposit Policy Statement.* Retrieved June 16, 2025, from https://nrcrecycles.org/nrc-container-deposit-policy-statement/.

57. Congress.gov. (n.d.). *Home.* Retrieved June 16, 2025, from https://www.congress.gov/.

58. Parliament of Canada. (n.d.). *Welcome to the House of Commons.* Retrieved June 16, 2025, from https://www.ourcommons.ca/en.

EPILOGUE

1. Lincoln, A. (1865, March 4). *Lincoln's Second Inaugural Address.* National Park Service—Lincoln Memorial. https://www.nps.gov/linc/learn/historyculture/lincoln-second-inaugural.htm.

2. *Midrash Ecclesiastes Rabbah 7:13* (AJWS Staff, Trans.). (n.d.). Sefaria. https://www.sefaria.org/profile/ajws-staff?tab=sheets.

3. Cascio, J. (2009, September 28). The Next Big Thing: Resilience. *Foreign Policy.* https://foreignpolicy.com/2009/09/28/the-next-big-thing-resilience/.

4. Baptista, A. (2016, November 13). What Will Change: Ana Baptista on Environmental Justice. *Tishman Environment and Design Center Blog.* https://www.tishmancenter.org/blog/what-will-change-ana-baptista.

5. Ibid.

6. Kingston, M. H. (2003, September). *The Fifth Book of Peace.* Spirituality & Practice. https://www.spiritualityandpractice.com/books/reviews/view/6803?id=6803.

7. Kolbert, E. 2014. *The Sixth Extinction: An Unnatural History.* Henry Holt and Company.

8. Siewert, C. (2022, April 4). Consciousness and Intentionality. In Zalta, E., N. (Ed.), *The Stanford Encyclopedia of Philosophy* (Summer 2022 edition). https://plato.stanford.edu/archives/sum2022/entries/consciousness-intentionality.

9. McDonough, W. (2005, July 14). In Twist, J., Eco-designs on future cities. *BBC News.* http://news.bbc.co.uk/2/hi/science/nature/4682011.stm.

10. Intergovernmental Panel on Climate Change (IPCC). (2018). *Special Report: Global Warming of 1.5° C: Strengthening and Implementing the Global Response.* https://www.ipcc.ch/sr15/chapter/chapter-4/.

11. Brooks, D. (2023). *How to Know a Person: The Art of Seeing Others Deeply and Being Deeply Seen*. Penguin Random House.

12. Gershon, L. (2015, September 29). Consumerism: An Economic Critique. *JSTOR Daily*. https://daily.jstor.org/consumerism-economic-critique/.

13. Quantis. (2023, April 4). Moving Beyond Rhetoric: To Build a Nature-Positive Economy, Focus on Nature-Supportive Action. *Quantis Insights*. https://quantis.com/insights/defining-nature-positive-business/.

14. Quinn, M. (2024, December 3). Kansas County Sues Major Plastic Producers, Alleging Deceptive Recycling Messaging. *WasteDive*. https://www.wastedive.com/news/kansas-lawsuit-plastics-producers-american-chemistry-council-exxonmobil-eastman/734443/.

15. UN Environment Programme. (n.d.). *Intergovernmental Negotiating Committee on Plastic Pollution*. Retrieved June 16, 2025, from https://www.unep.org/inc-plastic-pollution/.

16. Smieja, J. (2024, December 5). Plastics Treaty Negotiations Remain Unfinished. What's Next? *Trellis*. https://trellis.net/article/plastics-treaty-negotiations-remain-unfinished-whats-next/.

17. Jeong, A. (2024, December 1). Divided over Whether to Stop Making Plastic, U.N. Treaty Talks Collapse. *The Washington Post*. https://www.washingtonpost.com/climate-environment/2024/12/01/plastic-pollution-treaty-global-un-busan/.

18. Laville, S. (2024, November 24). Plastics Lobbyists Make Up Biggest Group at Vital UN Treaty Talks. *The Guardian*. https://www.theguardian.com/environment/2024/nov/27/plastic-lobbyists-biggest-group-un-treaty-talks-busan-korea.

19. Pazzanese, C. (2019, May 30). Merkel Advises Graduates: Break the Walls That Hem You In. *The Harvard Gazette*. https://news.harvard.edu/gazette/story/2019/05/at-harvard-commencement-merkel-tells-grads-break-the-walls-that-hem-you-in/.

20. Ibid.

21. Thunberg, G. (2019, December 11). In Beament, E., Greta Tells COP25: "This Has to Stop." *The Ecologist*. https://theecologist.org/2019/dec/11/greta-tells-cop25-has-stop?utm_source=chatgpt.com.

22. Thunberg, G. (2021, September). In Carrington, D., "Blah, Blah, Blah": Greta Thunberg Lambasts Leaders over Climate Crisis. *The Guardian*. https://www.theguardian.com/environment/2021/sep/28/blah-greta-thunberg-leaders-climate-crisis-co2-emissions?utm_source=chatgpt.com.

23. Borenstein, S., Keyton, D., Keaten, J., & Arasu, S. (2023, December 13). In a First, Delegates at UN Climate Talks Agree to Transition Away from Planet-Warming Fossil Fuels *AP News*. https://apnews.com/article/cop28-climate-summit-negotiations-fossil-fuels-dubai-64c0e39e6ad54a98e05e5201a2215293.

APPENDIX

1. United Nations Climate Change. (2021, October 5). *World Religious Leaders and Scientists Make Pre-COP26 Appeal.* https://unfccc.int/news/world-religious-leaders-and-scientists-make-pre-cop26-appeal.

2. Shaw, M. D. (2003). *Thich Nhat Hanh: Buddhism in Action* (Quote by Thich Nhat Hanh). SkyLight Paths Publishing.

3. Earth Charter Comission. (2000). *The Earth Charter.* https://earthcharter.org/read-the-earth-charter/.

4. Intergovernmental Negotiation Committee on Plastic Pollution. (2024). *Bridge to Busan and Beyond: Declaration on Primary Plastic Polymers.* https://www.bridgetobusan.com/ppp.

ABOUT THE AUTHOR

BOB GEDERT is a Cincinnati-based recycling consultant with forty-five years of experience in guiding communities toward sustainable materials management through a systems approach, bridging recycling best practices and sustainability toward local circular economies. He has experience in building local economic development practices with zero-waste goals through private-public partnerships that utilize local markets for recyclables through innovative entrepreneurship and collaborative business networks.

Previously, Bob served as department director of Austin Resource Recovery, leading the city toward zero-waste goals. In addition to his tenure in Austin, he has served in the leadership of resource recovery in many communities including the city of Fresno, California, as chief of recycling operations; the counties of Auglaize and Highland in Ohio; the city of Cincinnati, Ohio; and communities across Indiana as chief of source reduction and recycling with the Indiana Department of Environmental Management.

Bob is the former president of the National Recycling Coalition and previously served on the board of directors for the Indiana Recycling Coalition and the Association of Ohio Recyclers. He was also the executive director of the California Resource Recovery Association and a founding member of the California Product Stewardship Council and the National Stewardship Action Council.

In 2019, Bob was awarded the Lifetime Recycling Achievement Award from the National Recycling Coalition. In 2024, Bob was selected for inclusion in Marquis Who's Who for Expertise in Education and Waste Management.

Bob is an adjunct professor at Xavier University in Cincinnati, teaching climate change. He is married to his college sweetheart, Kathy. His greatest pleasure in life is watching his three children and two grandchildren make a difference in their communities. He enjoys walking in the local parks with his wife.

ABOUT THE ARTISTS

PAM LONGOBARDI

Pam Longobardi's parents, an ocean lifeguard and Delaware's female diving and swimming champion, connected her from an early age to water life. She moved to Atlanta in 1970 and saw her beloved neighborhood pond drained to build the high school she attended. Since then, she lived for varying time periods in Wyoming, Montana, California, and Tennessee and worked as a firefighter and tree planter, a scientific illustrator and an aerial mapmaker, a collaborative fine art edition printer and color mixer, and a designer and freelance artist. Discovering mountains of plastic on remote Hawai'ian shores in 2005, she founded the Drifters Project in 2006, centralizing the artist as culture worker/activist/researcher. Now a global collaborative entity, Drifters Project has removed tens of thousands of pounds of material from natural environments and re-situated it as social sculpture. Her studio-based and social practice ranges from paintings, collage, photography, large-scale sculpture, installation to public actions and performance and has been exhibited around the world. Longobardi has shown her artwork across the U.S. and in Greece, Monaco, Germany, Finland, Wales, Britain, Slovakia, China, Japan, Italy, Spain, Belgium, Poland and twice done projects as part of the Venice Biennale. Winner of the prestigious Hudgens Prize, Longobardi was featured in *National Geographic*, *Sierra* magazine, Weather Channel, multiple films, and public television and news broadcasts. As Oceanic Society's Artist-In-Nature, she co-leads expeditions to remote places, working with participant communities remediating plastic and its environmental impact, making public art, and fostering behavior change. She lives in Atlanta, Georgia, as regent's professor at Georgia State University. A sixteen-year survey of her work opened in January 2022 at the Baker Museum in Naples, Florida, with a comprehensive monograph published by Fall Line Press titled *Ocean Gleaning*. Longobardi's work is framed within a conversation about globalism and conservation.

KALLIOPI MONOYIOS

Kalliopi Monoyios is an artist working to mature the conversation around plastic and American consumerism. She uses art to bring people back to a state of childlike inquiry, where we can see our collective behavior and societal decisions with new eyes. She takes workhorse plastic items that are largely overlooked and undervalued—dental floss, plastic packaging, contact lenses, and computer cords—and pairs them with similarly undervalued and traditionally un- or underpaid skills: sewing, embroidery, weaving, quilting. By spending hours manipulating objects that have been deemed trash, she poses critical questions about what we value and why.